I0054170

DEBORAH WHEELER

SILK CLOUDS AND OLIVE TREES

Stories from the Battle of Crete

Author: Deborah C Wheeler

ABN: 8914 9961 670

Website: www.deborahcwheeler.com

Email: read@deborahcwheeler.com

Copyright © 2021

First Published May 2021

All care and due diligence has been taken in compiling the information on each soldier listed in this publication, as well as the war record of each person. Not all information found has been included and we apologise if some information listed is found to be incorrect. Thanks to Alfred Carpenter and the families of the soldiers for their valuable contributions.

The moral rights of the author have been asserted.

All rights reserved.

This book may not be reproduced in whole or in part, stored, posted on the internet or transmitted in any form or by any means, whether electronically, mechanically, or by photocopying, recording, sharing or any other means, without written permission from the author and publisher of the book. Please feel free to email me for permission – I'm usually obliging. All content found on or offline without written permission from me will be breaking the copyright law and therefore render you liable and at risk of persecution.

ISBN: 978-0-6481091-6-7

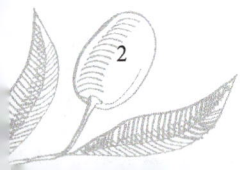

DEDICATION

To the brave and honourable people of Greece and Crete who offered everything they had to help our ANZACs during the Second World War.

Sometimes this friendship came at a great cost, including the ultimate sacrifice of their own lives and the lives of their families and friends.

For the Cretans it was not a ten-day battle. Their battle lasted from 20 May 1941 until the end of the war some four years later in 1945. Their sacrifices guaranteed the survival of many of their Allies to fight another day.

Gone but not forgotten as the strong bonds of those friendships formed 80 years ago continue today, passed down from that generation to the next.

CONTENTS

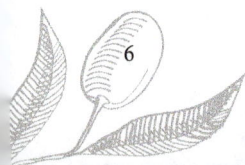

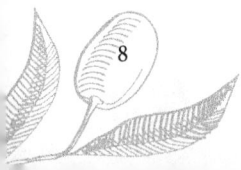

FOREWORD

by Alf Carpenter OAM, EO, JP, Battle of Crete Veteran

I am honoured to have been invited by Deborah C Wheeler to write the Foreword for her book, Silk Clouds and Olive Trees about the Second World War actions of some of the men involved in the Greek Offensive and the Battle of Crete.

I feel quite qualified to apply myself to this task, having served as a member of the 2/4th Australian Infantry Battalion as Regimental Sergeant Major, during the Greek and Crete actions.

In the book, Deborah makes mention of Sergeant Albert (Gus) William MAYER NX5125, a fellow member of my unit who was attached to the 2/4th Transport Paton. I remember Gus well.

I have checked events described in the book with my RSM Action Diary and I can verify the feelings of each soldier during these actions, especially those experienced as Stuka Dive Bombers with their sirens screaming came raining down in multiple numbers. They certainly made one's hair stand on end. My old Batman, Charlie Pearson would dive for cover under any ground depression, take a swig from his water bottle, usually charged with SRD rum or whisky so he could fortify himself enough to watch these planes. When questioned he would reply, "I want to be able to tell my grandchildren about the Dive Bombers in the future." He was a great Batman.

Deborah is to be congratulated on her detailed research and for writing a publication of Second World War actions by Australian and New Zealand Forces before memories disappear.

NX Alfred Clive CARPENTER
Ex RSM 2/4th Australian Infantry Battalion
Adjutant 42nd Land Craft Company
DAQMG Australian Staff Corps.
Darwin Northern Territory Force
A Carpenter OAM, EO, JP.

PREFACE

Discovering 80 years of unshakeable friendship

It was a cool, clear winter's morning with the sun shining brightly in 2018 when I left Warwick to go to Stanthorpe, a town rich in stone fruit, abundant wineries and as the town sits on the highest peak in Queensland, it's also known for its extremely cold weather. Travelling back to the area which had been home to me on and off for forty years was always an enjoyable experience. It was a great opportunity to do some work and catch up with old friends for lunch. On this particular day I was off to discuss volume two of my upcoming book with the secretary of the RSL Sub-Branch.

The Sub-Branch had commissioned me to undertake research on the Kyoomba Sanatorium, a medical facility used by the AIF during and after the First World War for return soldiers suffering from tuberculosis. Nothing had ever been written or recorded in any detail about the place or the people who had been there. It was evident that I would have a huge task ahead of me.

Never having undertaken a work of this magnitude I soon became totally captivated by the whole process of in depth research and project construction. Right from the initial searches, I realised how important it would be to unearth as many of the soldiers, doctors, nurses and workers connected to this long forgotten military medical facility as possible.

It's a bit like people who look into their family history; once you begin, it's hard to stop—the more you learn, the more you're captivated. Interesting stories and connections pop out of the woodwork, emerging and taking on a life of their own as they reach out to be heard. This, as yet unknown project, about the Battle of Crete would be no different for me.

Driving into Applethorpe on that cool winter's morning I saw a man walking along the roadside carrying two very large flags and heading in the direction of Warwick. 'Curious,' I thought. For some reason I knew it was important for me to stop and speak with him.

You could not have had a more unlikely meeting. We stood off to the side of the road and introduced ourselves. He identified himself as Andreas Lionakis. "Why are you doing this walk?" I asked. He replied the walk was to honour his father's memory and to raise awareness of the Battle of Crete. He briefly told me the story of how his late father had been involved in the battle as a sixteen year old fighting with the resistance. How his father had always spoken with pride of the ANZACs who fought alongside the Cretans during this battle. His father told him the ANZACs were the bravest and most honourable men he had ever met. Andreas, who was born in Australia, was also looking to reconnect with the Greek side of his family heritage.

Because of my work on the long forgotten Kyoomba Sanatorium, I was intrigued and had a genuine desire to learn more about the importance of this little recognised Second World War battle.

Unfortunately our brief meeting had to end as we both needed to go to our respective destinations. I asked if he had accommodation in Warwick. He had one night booked but was looking to stay another night. I gave him my business card and told him there would be a bed and dinner waiting for him, gratis, when he eventually arrived in Warwick.

Almost a week later, Andreas arrived at our home. We sat over a cuppa with some freshly baked scones and settled down for a good long chin wag. This man and his quest were intriguing and I wanted to find out more about his captivating history.

Initially when looking to do something he thought, I don't run or ride push bikes, so I'll just walk. The significance of walking from Brisbane to Wallangarra soon became apparent. When Andreas was doing his research he found this distance was the same distance as the length of Crete. Also wanting to do something worthwhile for today's military he decided to raise money for Mates4Mates, a veteran's organisation dedicated to helping current and retired veterans and their families. So began the start of what was to be an annual event for Andreas and a much-valued fundraising event for Mates4Mates.

I knew absolutely nothing about the Battle of Crete but Andreas willingly shared his knowledge. He told me how during the Second World War the small island of Crete provided landing fields for aeroplanes and harbours for shipping and was a stronghold for the Allies. How the Cretans fought alongside the British and the ANZACs to defend their island.

His passion and enthusiasm to have the Battle of Crete recognised, his desire to continue to cultivate ties between Australia, New Zealand and Cretan people, combined with his wish to honour his father, Tim Lionakis, stirred me to the core.

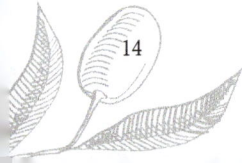

In 2019 Andreas once again arrived to stay with us for two nights. I had organised a meeting where I introduced Andreas to the then Mayor of the Southern Downs. As a retired veteran herself, the mayor was very keen to work with us to continue to forge friendships between Australia and Crete.

Andreas, former Mayor Tracy Dobie & Deb
Photo: Courtesy of Southern Downs Regional Council

LEST WE FORGET

Andreas had no expectations of meeting anyone with a connection to the Battle of Crete. However, fate was on Andreas' side as a number of people with relatives who fought in the Battle of Crete reached out and made contact with him.

Three days into his first walk, Andreas realised why he was really doing it. "I was tired, sore and had blisters on my feet. I came across two men who were fixing a fence. I asked if there was any accommodation nearby. They told me I must be talking about Harrisville. It's just up the road. I ended up in Harrisville totally off track. Was I happy to arrive at the Royal Hotel. While I was sitting having a beer, people saw the flags and started asking what I was doing, so I told them my story. A gentleman in the corner of the bar told me he had come across his grandfather's military documents. His grandfather had fought in the Battle of Crete. That was the moment I understood the walk I was doing was what I was meant to do."

Hearing this part of Andreas' story was a light bulb moment for me. Instantly I felt compelled to connect with the descendants of these men. It was the individual stories I found absolutely fascinating. Uncovering and sharing stories of the ANZACs with connections to my local region was an important and undiscovered piece of the fabric of our local military history. This was something special.

While gathering information on those men, I also uncovered connecting stories about Alf Carpenter, Clifford Morris, Dimitri Zampelis and Englishman John Slack.

I would like to personally recognise Maggie Paxton-Smith for her generosity in sending me the booklet, *Alf Carpenter: Second to none*. Maggie lovingly prepared this booklet in 2017 for Alf's 100th birthday celebrations.

The lives of these men all connected because of the Greek Offensive, the ensuing Battle of Crete and in some cases their shared POW experiences.

The Greek Offensive and the Battle of Crete came about after Italy gained control of Albania. Italian dictator and Prime Minister Benito Mussolini, not wanting to be upstaged by the Germans, set his sights on a surprise attack on Greece.

The Italians invaded Greece on 28 October 1940. This was the start of Greece's involvement in the Second World War and within a month the Greek Army had successfully driven the Italians back into Albania. Mussolini's failure now forced Hitler to halt his preparations to invade the Soviet Union. The German invasion of Greece started on 6 April 1941.

Deborah C Wheeler

CHRONOLOGY

This chronology is not intended to be comprehensive. Rather it serves to give a brief overview chronicling events that occurred during Operation Lustre in Greece and Operation Mercury in Crete.

1940

~ Italy endeavours to invade Greece but is turned back by the Greek Army. This results in the Axis powers of Germany, Italy and Bulgaria combining forces and planning an invasion of Greece with a far stronger force of men, equipment and air support.

1941

~ Commonwealth forces join with Greece to fight invading Axis powers.

~ Allied troops embark for Greece.

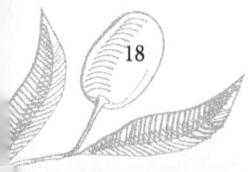

~ The Allies put Operation Lustre into action at Vevi Pass between the borders of Greece and Yugoslavia.

~ Axis powers defeat Yugoslavian troops and cross into Greece.

Superior numbers of troops, weaponry and air support make defence of Greece impossible.

~ Allied troops work their way down through Greece while continuing to maintain strong rear-guard action against the enemy. Execution of the withdrawal guarantees heavy losses for invading troops.

~ Success of the rear-guard action ensures thousands of Allied troops are safely withdrawn to Egypt and the island of Crete.

~ April 25: Hitler approves attack on Crete.

~ April 29: Main Allied evacuations from Greece end.

~ April 30: Greece officially falls to the Axis powers.

~ April 30: Major-General Bernard Freyberg appointed commander of the defence force in Crete.

~ May 16: Last British reinforcements arrive in Crete.

~ May 19: Freyberg orders the last airworthy aircraft to be flown out of Crete to Egypt leaving the island with no aerial defences.

~ May 20: Germany launches largest airborne attack ever used in warfare against Allies on Crete. Germany suffers appalling losses with approximately 4,000 of the 10,000 German elite paratroopers killed on first day of fighting without achieving any of their objectives.

~ May 21: Germany succeeds in taking control of airfield at Maleme after Allies suffer communication problems and counter attack was delayed.

~ May 22: Germans control airfield at Maleme and now are able
to safely land thousands of troops. Allies counter attack but are
forced to withdraw.

~ May 23: Germans consolidate position east of Maleme.

~ May 24: King of Greece and his government taken off Crete by two
British Destroyers. There was an acceptance that the island had been lost.

~ May 25: Continuous air raids all day. Germans take Galatas but
recaptured by New Zealand troops.

~ May 26: ANZACs form defensive line along Chania to Tsikalaria
road.

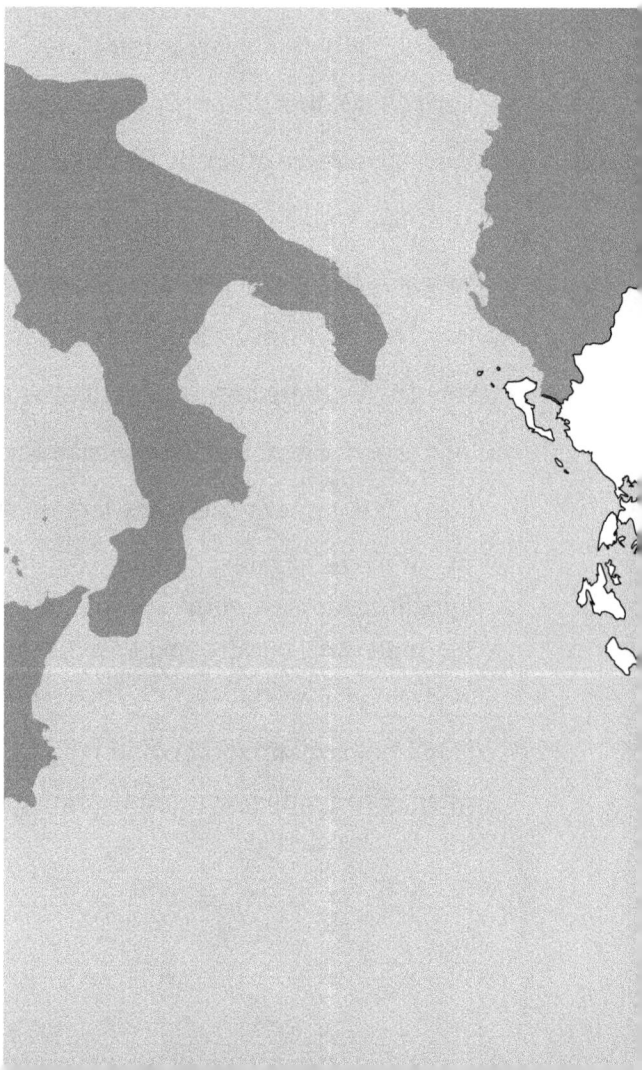

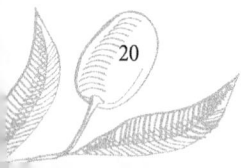

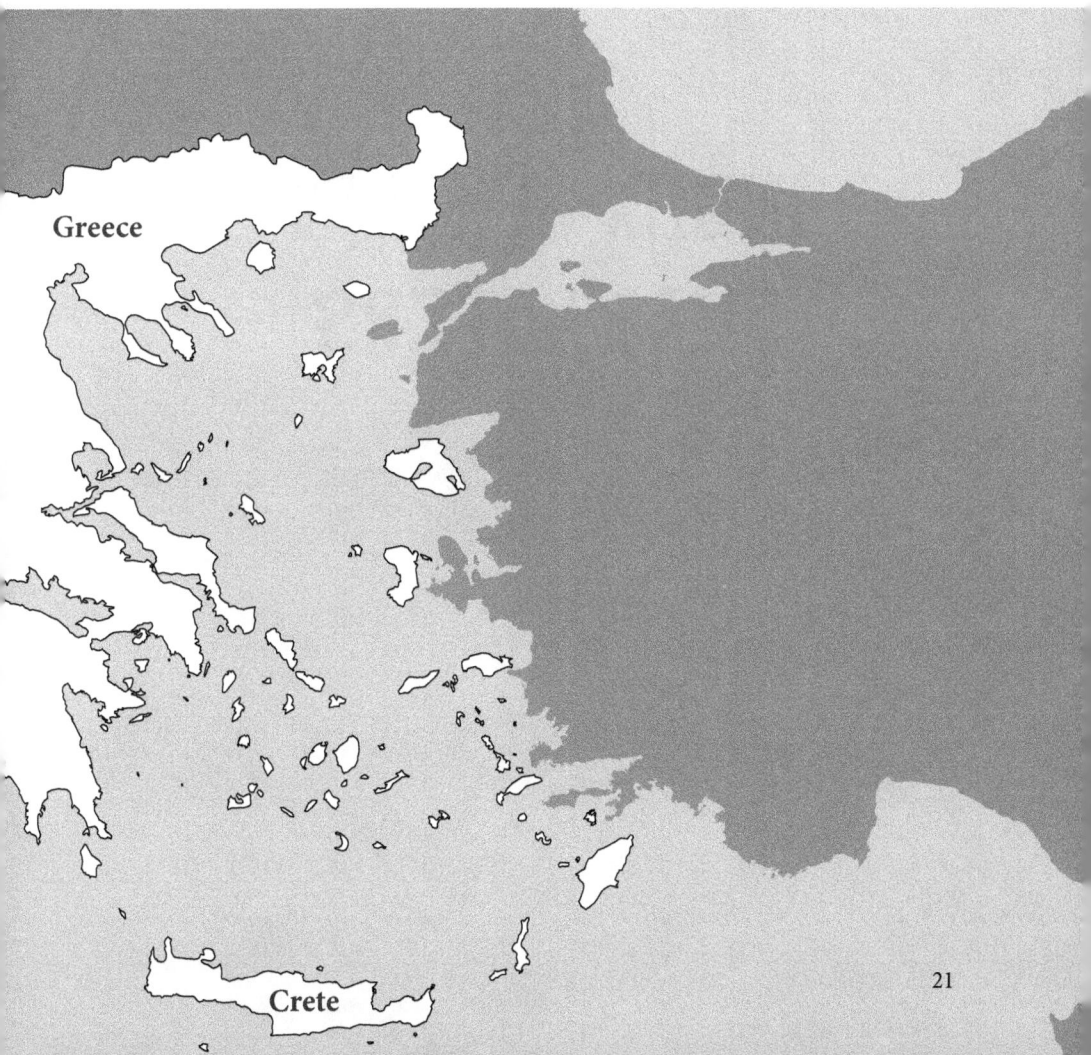

- ~ May 27: Freyberg receives authorisation to withdraw to Sphakia.

- ~ May 27: ANZACs form rear-guard at 42nd Street gaining precious time for Commonwealth force retreating southward.

- ~ May 28: Evacuations begin.

- ~ May 29: Allied troops traverse White Mountains, heading for Sphakia.

- ~ May 30: Allied troops at Retimo do not receive call to evacuate. All remaining troops taken as POWs

- ~ May 31: Germany takes control of the island.

Greece

Crete

INDEX OF MAIN CHARACTERS

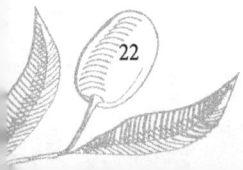

BORRIE, Andrew Findley

Service No: 10495
Unit: New Zealand Expeditionary Force
Rank: Driver
Place of Birth: New Zealand
Date of Birth: 1900
Date Enlisted: 1940
Age on Enlistment: 40
Place of Enlistment: New Zealand
Pre-War Occupation: Motor Driver
Date of Death: 1963
Place of Burial: Purewa Cemetery, Auckland

Falsified age down by five years in order to enlist. Served as a driver in 5th Field Ambulance. Involved in Operation Lustre and part of force withdrawn from Greece and taken to Crete. Captured and taken POW on Crete, interned at Stalag VIII-B Lamsdorf, Poland 1941; transferred to Oflag IX-A in 1942; again transferred to Stalag IV-A in 1943 before being repatriated to New Zealand after release. Cousin of Dr John Borrie MBE (POW) and Dr Alexander Borrie MC.

BORRIE, John MBE

Service No: 37751
Unit: New Zealand Medical Corps
Rank: Officer in Command (Doctor)
Place of Birth: Port Chalmers, Dunedin, New Zealand
Date of Birth: 22 January 1915
Date Enlisted: 1940
Age on Enlistment: 25
Place of Enlistment: New Zealand
Pre-War Occupation: Medical Practitioner
Date of Death: 1 August 2006
Place of Burial: Dunedin, New Zealand

Doctor, captured in Greece on 1 April 1941. During his capture he worked in prison hospitals in Greece, Upper Silesia and Stalag VIII-B Lamsdorf, Poland. After his release from the POW camp at the end of the war he was repatriated to United Kingdom where he worked in a number of hospitals before returning to New Zealand. Brother of Major Alex Borrie MC. Alex awarded Military Cross for rescuing New Zealand wounded at Monte Cassino in 1943. Cousin of Andrew Findley Borrie (POW).

CARPENTER, Alfred [Alf] Clive OAM, EO, JP

Service No: NX5979
Unit: 2/4th Australian Infantry Battalion
Rank: Regimental Sergeant Major
Place of Birth: Wagga Wagga, NSW
Date of Birth: 22 April 1917
Date Enlisted: 3 November 1939
Age on Enlistment: 22
Place of Enlistment: Sydney, NSW
Pre-War Occupation: Shop assistant
Date of Death: Not applicable. Alf is still alive and doing well in 2021.
Place of Burial: Not applicable in 2021

Prior to the war Alf worked at W.G. Huthwaite's hardware store while his good mate Charlie Jewell worked at 'Porky' Richard's delicatessen. On weekends he could be found working at the fire station or patrolling at the Wagga Wagga Beach Lifesaving Club. He was awarded the Royal Lifesaving Bronze and Silver medallions and the Second Class Instructor's Certificate. While in the Middle East Alf formed the 'Gaza Beach Lifesaving Patrol'. Alf and Charlie enlisted on the same day and were both assigned to the 2/4th Battalion. They served side by side in the Middle East and fought in the Battle of Tobruk. Alf is one of the few 'Rats of Tobruk' still with us in 2021. They were deployed to Greece in 1941. During the withdrawal from Greece, Charlie was killed on 22 April 1941. Alf was devastated at the loss of his good mate. Evacuated to Crete. Escaped to Alexandria with Sgt Albert William Mayer [Gus]. Returned to Australia at the end of the war.

CHANNELL, Douglas Ronald MC

Service No: NX115
Unit: 2/1st Australian Infantry Battalion
Rank: Captain
Place of Birth: London, England
Date of Birth: 7 December 1909
Date Enlisted: 25 October 1939
Age on Enlistment: 29
Place of Enlistment: Sydney, NSW
Pre-War Occupation: Radio announcer
Date of Death: 25 May 1982
Place of Burial: Brisbane, Queensland

First member of the Australian Broadcasting Commission (ABC) staff to enlist. Served in the Middle East before being deployed to Greece to join in Operation Lustre. Awarded the Military Cross for his actions at Retimo on Hill A in Crete. Seriously wounded and taken prisoner at Retimo. Sent to Oflag VII-B Colditz. Set free from POW camp at end of war and repatriated to England. Worked for BBC before returning to Australia and going back to work for the ABC. Attended the Investiture at Government House Sydney on the 6 December 1949, where he was awarded the Military Cross by the governor, Lieutenant General Northcott.

COLLINS, Ronald (Splinter) Alexander

Service No: QX1248
Unit: 2/1st Australian Anti-Tank Regiment
Rank: Sapper
Place of Birth: Brisbane, Queensland
Date of Birth: 01 February 1918
Date Enlisted: 3 November 1939
Age on Enlistment: 21
Place of Enlistment: Gatton College, QLD
Pre-War Occupation: Agricultural Student
Date of Death: 21 February 2013
Place of Burial: Interred at Garden of Remembrance, Pinnaroo, Brisbane

Initially assigned to the Royal Australian Engineers unit before being transferred to the 2nd/1st Australian Anti-Tank Regiment. Fought in Operation Lustre in Greece. Evacuated to Crete with good mate Glen Scott. Escaped from Crete the day before Operation Mercury commenced. Disembarked in Alexandria. Returned to Australia, disembarking in Adelaide. Posted to Thursday Island, Wagga Wagga and spent time building bridges on the Murrumbidgee River before returning to Brisbane. Ron's brother Henry William Collins (QX14618) was captured by the Japanese and taken as a POW on Borneo. Henry died of malnutrition while in captivity in March 1945.

EASON, Alan

Service No: QX17451
Unit: 1Corps. Troop Supply
Rank: Private
Place of Birth: Brisbane, QLD
Date of Birth: 12 December 1917
Date Enlisted: 29 July 1940
Age on Enlistment: 22
Place of Enlistment: Kelvin Grove, QLD
Pre-War Occupation: Commercial traveller
Date of Death: 20 February 1945
Place of Burial: Klagenfurt War Cemetery, Klagenfurt, Austria

Alan was a super keen footballer representing not only his Brisbane club but also his state as a member of the Queensland Rugby Union Touring Team of 1939. Alan and his three brothers, Raymond Cyril Eason (QX39435), Colin Leonard Eason (Q126659) and Keith Alexander Eason (QX6780), all enlisted in the Second World War. Embarked for Greece as part of Operation Lustre before being evacuated to Crete after Germany forced the Allies to retreat and withdraw from Greece. Captured by Germans and sent as a POW to Stalag VXIII-A in Austria. Held as POW from May 1941 until his death on 20 February 1945. Accidentally killed by friendly fire in bombing raid. Buried at the Klagenfurt War Cemetery, Klagenfurt, Austria.

MAYER, Albert (Gus) William MID

Service No: NX5125
Unit: 2nd/4th Battalion
Rank: Sergeant
Place of Birth: Carlton, NSW
Date of Birth: 25 January 1918
Date Enlisted: 21 October 1939
Age on Enlistment: 21
Place of Enlistment: Parramatta, NSW
Pre-War Occupation: Salesman
Date of Death: 13 November 1968
Place of Burial: Rookwood, NSW

Sergeant Transport Unit. Gus served in the Middle East before being deployed to Greece. Evacuated to Crete. Escaped from Crete going to Alexandria with RSM Alfred [Alf] Carpenter. It is noted on his service file that on 2 May 1942 he was Mentioned in Dispatches. Unfortunately at the time of going to print the writer has been unable to discover the relevant citation. Gus went on to serve in the South West Pacific Area from 28 October 1944 until 9 August 1945. He was discharged on the 4 September 1945. Returned to Australia at the end of the war. Glen Derek SCOTT of the 2/1st Anti-Tank Regiment went on to become his brother-in-law after marrying Gus' sister, Weipa.

LIONAKIS, Tim

Place of Birth: Monessen, Pennsylvania,
United States of America
Date of Birth: 26 December 1924
Date of Death: 1 July 2014
Place of Burial: Mount Gravatt,
Queensland

Civilian – Resistance Fighter

Tim was born in the United States of America to a Cretan father. At the age of seven Tim and his younger sister were sent to his father's family in Crete after the death of their mother. Tim was just 16 when the Battle of Crete commenced. Tim became a member of the Cretan Resistance. After escaping from Crete to Athens, Tim worked as a civilian guard for the USA at the Eleusis Air Field in 1944. He was injured during a conflict and sustained severe shrapnel wounds to his left leg and two bullet wounds to his left arm. He was evacuated from Greece on a Dutch Red Cross ship heading for America. The doctor on board assessed the damage and told Tim he wanted to amputate the arm. Tim refused point blank. The nurses on board the ship treated his injuries and nursed him back to health. Despite Tim's courageous efforts fighting alongside the Allies, he was not recognised as an official soldier. Hence, once he recovered he was required to pay back his fare to the Red Cross for his passage. This one episode left a bitter taste in his mouth.

This was not to be the only time Tim faced adversity because of Government policy. However, Tim's belief in the ANZAC soldiers who fought in the battle of Crete never wavered as he always told his children how brave these soldiers were. After living in America for a number of years, Tim left and travelled around many different countries. He eventually migrated to Australia with his wife. Tim went on to become a very successful businessman in Central Queensland. His son Andreas does an annual walk from Brisbane to Wallangarra and back to commemorate his father and to continue to strengthen ties between Australians and Cretans. Andreas inspired Deborah Wheeler (author) to learn more about the Battle of Crete and to record the stories of a few Aussie and New Zealand soldiers whose families reached out to him during the course of his annual walks. Andreas acknowledges two very special people who assisted him in becoming the man he is today. Firstly, his late father, mentor and friend, Tim Lionakis. Secondly, his best friend Tina Hatzifotis, who has influenced his life in such a positive way.

MORRIS, Clifford [Cliff] Harrie

Service No: VX15068
Unit: 2/5th Australian Infantry Battalion
Rank: Private
Place of Birth: Mildura, VIC
Date of Birth: 4 October 1915
Date Enlisted: 1 March 1940
Age on Enlistment: 24
Place of Enlistment: Ballarat, VIC
Pre-War Occupation: Transport driver
Date of Death: 12 January 1990
Place of Burial: Skipton, Victoria

Deployed to Greece on 9 April 1941. Evacuated to Crete. Listed as Missing in Action. International Red Cross confirmed Cliff taken as POW and interned at Stalag XVIII-B Austria. Made a number of failed escape attempts. After the war he sponsored his German guard Koloman and his family to come to Australia. His son, David Koloman Morris, was named in his honour as a mark of respect and thanks for all this guard had done for Cliff during his four years of captivity.

SAUNDERS, Reginald (Reg) MBE

Service No: VX12843
Unit: 2/7th Battalion
Rank: Captain
Place of Birth: Purnim, VIC
Date of Birth: 7 August 1920
Date Enlisted: 24 April 1940
Age on Enlistment: 19
Place of Enlistment: Caulfield, VIC
Pre-War Occupation: Timber/dairy worker
Date of Death: 2 March 1990
Place of Burial: Canberra, ACT

Reg's father, Chris Saunders, fought in the First World War and named his son after William Reginald Rawlings MM. Reg loved sport and playing football, cricket and boxing. Deployed to Greece and Crete. Involved in the bayonet charge at '42nd Street'. Stranded on Crete he refused to surrender and managed to avoid capture for eleven months with the help of sympathetic Cretans. In May 1942, he escaped aboard a fishing trawler to Libya. Reg's younger brother Harry Saunders (VX18629) of the 2/14th Battalion was killed in Papua New Guinea in December 1942. Reg was the first Indigenous Australian to be commissioned as an officer in the Australian Army. Reg worked hard to break down racial stereotypes. Reg enlisted again for the Korean War. His battalion (3ARA) was awarded the United States of America's Distinguished Unit Citation. Awarded the MBE in 1971 and appointed to the Council of the Australian War Memorial in 1985.

SCOTT, Glen (Stewy) Derek

Enlisted under assumed name: Gordon Allan STUART
Service No: QX4245
Unit: 2nd/1st Australian Anti-Tank Regiment
Rank: Gunner
Place of Birth: Mackay, QLD
Date of Birth: 10 Aug 1918 (7 Nov 1921)
Date Enlisted: 22 September 1939
Age on Enlistment: 21 (17)
Place of Enlistment: Brisbane, QLD
Pre-War Occupation: Farmer
Date of Death: 31 January 2008
Place of Burial: Stanthorpe, QLD

Even though the campaigns on Greece and Crete were devastating for the Allies, it was here where the members of the 2nd/1st formed their strongest relationships with the locals, especially on Crete. This was the only time in Glen's military career where he knew his back was covered by someone other than one of his fellow soldiers in the unit. It was something Glen and the members of his unit never forgot. After Crete, Glen was deployed to Syria, Palestine and New Guinea. After the war Glen married Weipa Mayer, sister of Albert (Gus) William Mayer. They purchased a fruit farm on the Southern Downs.

ZAMPELIS, Dimitri James (Jim)

Service No: VX989
Unit: 2/2nd Australian Field Regiment
Rank: Gunner
Place of Birth: Melbourne, VIC
Date of Birth: 22 December 1912
Date Enlisted: 6 November 1939
Age on Enlistment: 26
Place of Enlistment: East St Kilda, VIC
Pre-War Occupation: Waiter
Date of Death: 24 May 1941
Place of Burial: Crete. Remembered Athens Memorial, Greece

Melbourne was home to many Greek families. Here one had the opportunity to work hard and make a good living. Even though Jim's marriage to Doris did not work out, they were the proud parents of a beautiful baby boy, Peter. When Jim left to go overseas with the Australian Army, Peter was named as Jim's sole beneficiary. Jim often thought of his son's little face and appreciated the photos his own father sent to him. Jim was killed at Suda Bay, Crete, but to this day his body has never been recovered. Jim's granddaughter Lisa Zampelis is hopeful that one day he will lie side by side with his fallen mates in the Suda Bay War Cemetery.

ΕΛΛΗΝΙΚΗ ΔΗΜΟΚΡΑΤΙΑ
ΝΟΜΑΡΧΙΑΚΗ ΑΥΤΟΔΙΟΙΚΗΣΗ
ΡΕΘΥΜΝΗΣ

Η ΜΑΧΗ
ΤΗΣ
ΚΡΗΤΗΣ
THE BATTLE
OF
CRETE
1941

ΑΠΟΝΕΜΟΥΜΕ
WE AWARD

Αναμνηστικό Μετάλλιο Τιμής
ένεκεν στους Βετεράνους μαχητές
της Μάχης της Κρήτης
για την συμμετοχή τους
τον Μάιο του 1941
στον αγώνα
του Ελληνικού Λαού
εναντίον των Ναζιστικών
στρατευμάτων Κατοχής.

The commemorative
medal of honour
as a matter of courtesy to
the Veteran soldiers
of the Battle of Crete
for their participation
in the struggle
of the Greek people against
the Nazi occupying forces
in May 1941.

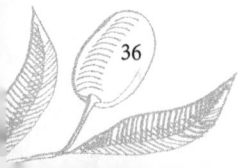

Ρέθυμνο 20 Μαΐου 2001
Ο Νομάρχης

Μανόλης Λίτινας

36

CHAPTER 1

The Englishman and the ABC Broadcast: Australia declares war on Germany

The Englishman

My name is Douglas Channell and my story started in England with my birth in 1907. Like most boys from that period, I attended school and joined the Territorial Army, achieving the rank of officer before leaving school. In 1921 I was too young to join the regular army so at the tender age of fourteen I decided to join the merchant navy. I had to lie about my age and told the captain I was fifteen. Being a big lad for my age, I could easily pass myself off as older.

I was after adventure and travel and I certainly got it. I experienced quite a few rough trips and the captain was not the best man to work for. After a couple of years the ship arrived in Townsville, Australia. By this stage the crew called a strike and refused to take orders from the rogue of a captain. I was one of the men sacked and left stranded in a country I knew little about, but on the other hand, I did like what I saw. At least here, unlike some of the other ports we had docked in, everyone spoke the Queen's English.

I had to borrow money and some shoes from one of the other seaman and set about looking for some work. In the newspaper there was a job going at the Stewart Creek Prison. The prison was ten miles (approximately 17 kilometres) out of Townsville, in what seemed to a young fellow like myself, the middle of nowhere. There was nothing around the area and nowhere to go. Applicants needed to be 21 years of age, so I had to lie about my age again as I was only 17 at the time. Anyway, I got the job and stayed for three years. There were very few prisoners at any one time and no escapes were attempted during my time there so life was pretty quiet. I did get a bit of a shock when a couple of sailors I had worked with on the boat ended up being incarcerated at the prison. Life here was too tame for me; the same monotonous procedures day in and day out. So while it was a steady job, I decided to move on.

Travelling around Queensland for a few years, I took work as a jackaroo. Can you imagine it, a big strapping Pommie like myself, standing 6 foot 3 inches in my socks riding around on a horse chasing cattle! I sensed the horse enjoyed throwing me to the ground on more occasions than I care to recount before I mastered the art of riding.

I soon got tired of the lonely life with too few people to talk and laugh with so I was off again. Back to Townsville where I secured a job as a salesman selling oil and tyres. Now, this was a little more like it. Different people to see every day and my gift of the gab ensured I made a good living. During this time I married but it was not meant to last.

I eventually worked my way down to Newcastle where I started work as an announcer for a commercial radio station.

Working for the ABC

When I originally auditioned for the Australian Broadcasting Commission (ABC), I was told I was 'too British'. By 1937, I was offered a job as their early morning announcer at station 2FC (ABC). This was more my style and something I was keen to undertake. I was in my element, especially when I convinced my boss to allow me to take the microphone out onto the street and conduct live interviews with the public. "The people will love it, an opportunity to hear themselves on the radio." We couldn't go wrong in my opinion and we didn't. 'Man in the Street' broadcasts became hugely successful and were a great addition to the ABC radio program. Because of the success of the show I was asked to also host 'Sydney Speaks'.

It was often reported that my program could put listeners in a good mood, sending them off to their jobs with a smile on their faces. Every morning people would gravitate to the wireless, keen to hear the morning's topic. Mothers would quite often get frustrated because they couldn't get the children to sit down to the breakfast table and many of the men would be caught spending too much time listening to the program and then have to run off to get to work.

Some of my more irrepressible remarks, which the listeners found so entertaining, were not always enjoyed by my boss. On more than one occasion, I was hauled before him and given a dressing down. But as my programs were so popular they continued on in the same vein much to the delight of the listeners. Life couldn't get any better; I was in my element.

On Saturday 2 September 1939, Sydney-siders were out in force to watch the rugby league premiership final between Balmain and South Sydney. The Sydney Cricket Ground had a staggering crowd of close to 27,000 footy fans all there to back their favourite team. After a second half score blitz Balmain were declared the premiership winners.

Footy was over and everyone was anticipating the warmer weather, relieved at being free of long cold winter days and nights. We all had so much to look forward to with the forthcoming start of the highly anticipated cricket season and tennis parties. A keen sportsman myself these were events I was looking forward to participating in. Life was good in Australia.

It was spring. Trees were coming into leaf, lawns were starting to grow, flowers were popping up in gardens and the stark landscape of winter was becoming tinged with green. Life would change dramatically the following day!

The broadcast

News from overseas over the last few months did not bode well for Europe. Newspaper headlines reported rumblings from Germany which had the whole of Europe on edge. What would Hitler and the Third Reich do next? We were soon to discover Hitler's plan when he attacked Poland and commenced mass bombardments of its cities at the beginning of September.

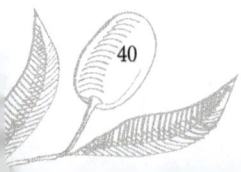

On Sunday 3 September 1939, my boss announced Prime Minister Robert Gordon Menzies would be making an address to the nation later in the evening. Little did I realise the impact of Britain's decision to declare war on Germany and how the upcoming speech would change the course of my life for the next six years.

As scheduled, at 9:15 pm, every national and commercial radio station announcer in Australia paused normal transmissions to tune into the Prime Minister's address to the nation. Those who owned radios had people from all over their neighbourhood visit. Everyone wanted to know what was about to be broadcast.

Prime Minister Menzies started his speech in a very sombre tone:

> *Fellow Australians, it is my melancholy duty to inform you officially that, in consequence of the persistence of Germany in her invasion of Poland, Great Britain has declared war upon her and that, as a result, Australia is also at war. No harder task can fall to the lot of a democratic leader than to make such an announcement. Great Britain and France, with the cooperation of the British Dominions, have struggled to avoid this tragedy. They have, as I firmly believe, been patient; they have kept the door of negotiation open; they have given no cause for aggression. But in the result their efforts have failed and we are, therefore, as a great family of nations, involved in a struggle which we must at all costs win, and which we believe in our hearts we will win ...*

> *It is plain - indeed it is brutally plain - that the Hitler ambition has been, not as he once said, to unite the German peoples under one rule, but to bring under that rule as many European countries, even of alien race, as can be subdued by force.*

If such a policy were allowed to go unchecked there could be no security in Europe and there could be no just peace for the world.

A halt has been called. Force has had to be resorted to check the march of force. Honest dealing, the peaceful adjustment of differences, the rights of independent peoples to live their own lives, the honouring of international obligations and promises - all these things are at stake.

There was never any doubt as to where Great Britain stood in relation to them. There can be no doubt that where Great Britain stands there stand the people of the entire British world.

Bitter as we all feel at this wanton crime, this is not a moment for rhetoric; prompt as the action of many thousands must be, it is for the rest a moment for quiet thinking; for that calm fortitude which rests not upon the beating of drums, but upon the unconquerable spirit of man, created by God in His own image. What may be before us we do not know, nor how long the journey. But this we do know, that Truth is our companion on that journey; that Truth is with us in the battle and that Truth must win.

Before I end, may I say this to you? In the bitter months that are to come, calmness, resoluteness, confidence and hard work will be required as never before. This war will involve not only soldiers and sailors and airmen, but supplies, foodstuffs, money. Our staying power, and particularly the staying power of the mother country, will be best assisted by keeping our production going; by continuing our avocations and our business as fully as we can; by maintaining employment and with it our strength.

I know that, in spite of the emotions we are all feeling, you will show that Australia is ready to see it through. May God, in His mercy and compassion, grant that the world may soon be delivered from this agony.

Prime Minister Robert G. Menzies: wartime broadcast. Source: AWM

The Courier-Mail

GLASSES

WANTED WATCHMAKERS

BRITAIN DECLARES WAR
AGAINST GERMANY OVER POLAND

MR. MENZIES' MESSAGE

FRANCE ALSO ACTS

Australia With Britain, Says Mr Menzies

Hitler Ignores Ultimatum

Britain yesterday declared herself at war with Germany. Last night, the Australian Prime Minister (Mr. Menzies), in a national broadcast, declared that Australia also was at war.

"I cannot believe that anything more, anything different could have been done."—Mr. Chamberlain.

"BRITAIN DECLARES WAR"
The Courier-Mail

Australians couldn't believe what they were hearing. Here we were again, only 20 years after the end of the Great War, touted as the war to end all wars, committed once again to sending our men and women off to fight in a war overseas. The country was just starting to get on its feet again after the Great Depression. People were only just starting to pull themselves together and now the carpet was to be ripped out from underneath them once again.

Mothers would again be called upon to compose themselves as they struggled to come to terms with what was happening in the world and the consequences it would have on their families. For many it would entail a twofold response. Some of the men who fought in the First World War looked to re-enlist. Fathers wanted to join up and stand side by side with their sons on the battlefield. Young men were keen to 'travel overseas' and do their bit for Australia. Daughters were also being asked to enlist in many roles, not only nursing. Politicians were beseeching the mothers of Australia to make a huge sacrifice.

These women would have little say in the final outcome. In some cases all the men in the household enlisted, leaving their women without any support, all in the name of doing the right thing and doing their 'duty'.

War! What a terrible waste for these women. Many would never see their fathers, husbands, brothers, sons or daughters again. These women and the remaining family at home with them would end up facing the same situation experienced by their mothers and grandmothers twenty years earlier.

Life's choices

For me the decision to enlist was an easy one. I was going to be a part of the 20,000 strong Expeditionary Force to be sent overseas as part of the newly formed Second AIF. I carried out my 'Man in the Street' session the following morning and when I signed off I walked out of the building and straight into the enlistment office around the corner. I later discovered this made me the first, but by no means the last, ABC employee to enlist.

Major General Kenneth William Eather was put in command of the newly raised 2/1st Infantry Battalion. He sourced recruits from the Militia in Sydney and I was one of the men chosen to be a member of the battalion.

I was assigned to the 2nd Australian Imperial Force, issued with Service Number NX115 and given the rank of lieutenant in the 2nd/1st Australian Infantry Battalion. Our basic training was undertaken at the Ingleburn Army Camp. In January 1940, we marched through Sydney to much fanfare and embarked for Palestine on the P&O ocean liner *SS Oxford*. Eather was a champion for mobile warfare in preference to the old style trench warfare and so our new training continued in Palestine. He was very much the disciplinarian and was known for handing down sentences of 28 days confinement to barracks for anyone who disobeyed his orders. Hence he acquired the nickname of 'February'. We went on to become lifelong friends.

CHAPTER 2

Mrs Stuart's little boy & Splinter

Scott aka Stuart

(Glen (Stewy) Scott)

At only 17 years and 10 months of age my parents would not give consent for me to enlist. Desperate to join the army, I left Cunnamulla in south-western Queensland where I had been working as a jackaroo on a large outback property. I loved the outdoor life and working with cattle, but now there was a war happening and I wanted to be a part of it. Once I made my mind up to enlist, nothing was going to stop me. After saying farewell to the property owners I hitchhiked all the way to Brisbane. The boss's wife had loaded me up with freshly baked biscuits and sandwiches. Thick slices of homemade bread, butter, corned beef and pickles.

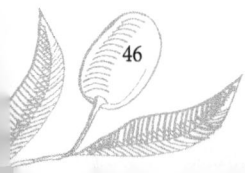

She wasn't about to see me starve as I headed off for Brisbane. My boss dropped me off in town and it wasn't long before I caught my first lift. The trip was over 800 miles (1290 kilometres) and it took me a couple of days to get there. Most people were pretty good and I had no difficulty getting lifts. I was a young man with his swag heading for Brisbane to join the army.

I picked up my last ride just outside Ipswich. A big red farm truck loaded with cattle pulled up in front of me. The old farmer asked me where I was headed, so I told him I was on my way to join the army and needed to find out where I had to go in order to enlist. He told me to jump on up as he was headed in that direction. He was a good man and we chatted about my time spent in Cunnamulla. He had been to Esk to pick up some young breeders and was heading home to his mixed crop farm in the Samford Valley. How lucky could I be. He dropped me off right outside the recruitment office at Kelvin Grove. He lifted his hand, waved farewell and wished me all the best as he drove off.

It made no difference that it was a Saturday, the recruitment office was open and they were accepting volunteers. So on 21 October 1939, I went in and signed my papers. Because I was underage I enlisted under the assumed name of Allan Gordon Stuart. After working my way through the enlistment paperwork and medical process I was issued with Service Number QX4245, put into the 1st Australian Anti-Tank Regiment and given the rank of gunner.

NOTE: I served under my false name, Allan G Stuart until 20 April 1943 when I signed a Statutory Declaration admitting my correct name was Glen Derek Scott, so my records could be correctly identified. My nickname 'Stewy' stayed with me for life.

In the army now!

I didn't need to report back until the Tuesday, so my last three days as a civilian were spent enjoying myself in Brisbane. It was grand being in the big smoke after the isolation of the bush and I took full advantage of the sights and sounds of the city. There w ere s o m any p eople, movies to watch, bars to visit (even though I was underage) and cafes to enjoy. On the Tuesday morning I reported back and was put into a group with other volunteers just like me. Orders were handed down instructing us to be ready for transportation by truck to an army camp where we would commence our basic training. Everything moved rather quickly from there. Upon arrival we jumped out of the truck and marched into Redbank Camp No. 4 Depot Company. For better or worse I was in the army now!

The first meal I had was by far the most disgusting I had ever eaten. Don't ask me what it was meant to be but it looked green and tasted awful. It was not surprising that the constant supply of tainted food resulted in many of the men in camp suffering from bouts of diarrhoea.

Forced route marches helped to break in the ill-fitting boots we were issued with. This was followed by the mandatory line up for our TA + B and Tetanus injections. Half of us got the needle in our arms while the other half received their injections in their stomachs. Straight after this we were sent off on another route march to work the injections through our systems. It didn't take long for the chaos to start. Most of the men who received the stomach jab doubled up with cramps or worse. Thankfully, I had received the jab in my arm.

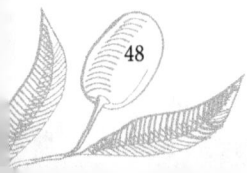

Following our basic training we headed south. I spent a short time at Liverpool, New South Wales, where we received our small pox injections using the same system used previously followed by the mandatory route march. Once again the results were the same, absolute chaos! You would swear they were using us as guinea pigs.

Early one evening while I was on picket duty at Central Railway Station, a very good-looking girl caught my eye. It was obvious from the way she was dressed she had just finished work and was walking down onto the railway platform ready to catch a train home. A rowdy group of soldiers followed her and started chatting her up. Casually, I strolled over to the blokes and asked the young lady if the soldiers were bothering her. I soon moved the guys on and stayed and chatted with her until her train arrived. Her name was Weipa and she asked me if I would like to go to her place for lunch on Sunday as her parents were heavily involved in the Canteen Fund at that time.

She told me her family invited soldiers from out of state to their home for lunch. This had become a common practice with so many men enlisting and being stationed away from home. The idea was to give new recruits a sense of family even though we were separated from our loved ones. Knowing I would be off duty on Sunday, I quickly accepted her invitation to lunch.

Sunday dawned with clear sunny skies. Setting off to the train station I wondered how the day would unfold. The house had a long driveway; I drew a deep breath and headed up the path to the steps. I had no idea what to expect. The beautiful big house was surrounded by lush gardens. Masses of rose bushes were in full bloom and the scent was just divine. It reminded me of Mum's roses back home.

Weipa's father greeted me at the door, shook my hand and welcomed me into their home. As he led me inside a female voice was calling out to Weipa, informing her of my arrival. Weipa descended the staircase dressed in a brightly coloured sun frock and gave me the biggest smile. I was introduced to the other members of her family and a couple of other diggers who had also been invited to join the family for lunch. Weipa's mum, Mrs Mayer, had set a wonderful spread in front of us and we all thoroughly enjoyed a delicious lunch and some good conversation.

So, that's how I came to meet Weipa Mayer, and boy I realised what a lucky bugger I was. She was the most beautiful girl and she really did turn my head. Beautiful, blonde and oh so sweet! I could have fallen in love right then and there. She promised to write to me so I gave her my army address. Her brother, Albert William Mayer (Gus) had already joined the army and was assigned to the 2/4th Australian Infantry Battalion.

After Liverpool we were sent to Ingleburn. Here one of the compulsory things we had to do was line up for the regulation 'short arm' inspection. We all thought this wouldn't be so bad really, especially not after the debacle with the injections. We were wrong. This little exercise became highly embarrassing, not just for me as a teenager in 1940 but for many of the older men, when the doctor turned out to be female. On reflection, it would seem between the food and the female doctor, the army was set to either experiment on or embarrass the men.

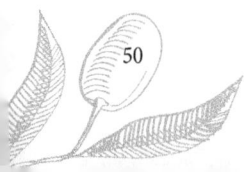

Over the blue sea

Eventually we were shipped overseas to Arena Road, Tidworth, England. It was here I became infected with a bad case of mumps and was placed in the isolation tent in camp with only an R.A.P. attendant to administer what help he could. Those of us in the tent were required to get our own sustenance, which proved harmful as we later found out, and resulted in a further two weeks in isolation at Tidworth Hospital. All up a very drawn out, painful experience, the side effects of which could have been avoided if only we had received the proper medical treatment in the first place.

Our next move was on to Colchester Camp where I was assigned the duty of dispatch rider. Outfitted with a new uniform and motorbike, I received instructions on my new responsibilities as a messenger, which included delivering messages and instructions to different units and commanding officers.

As we always seemed to move at night it was inevitable a gun tractor crew would become lost. On one occasion two tractors got lost and riding with only blackout headlights it was hard to find them in the pitch black. After successfully returning the tractors to the right track I was unfortunate to hit a cow crossing the road. Luckily for both the cow and myself I was not riding very fast. The crash resulted in a fall, denting both the motorbike and myself while the cow continued on its way. Pushing the bike all the way back to camp it was light before I eventually rejoined the unit. Boy, was I buggered!

At the height of 'The Blitz' we participated in constant anti-parachute patrols. We really knew we were in the war now. The destruction all around us was disconcerting. The attacks by the Germans on English soil only strengthened our resolve to fight to win the war as the last thing we wanted was for this to happen in Australia. I remember being shoulder to shoulder on the beach awaiting the invasion of the German army and forever thankful they did not come by sea.

From Colchester, the 2/4th Motor Transport Unit was sent to Swansea, Wales for transportation to the Middle East. Excitement reigned in the unit for goodness only knows what was awaiting us there.

Sweetheart

Below is a copy of a letter I wrote to Weipa. Despite my other love interests at the time, I thought maybe she could be the one I would eventually give my heart to, but first there was a war to win. This letter was written while on board the ship from Wales to Egypt.

Weipa

> *24th November 1940*
>
> *Dear Weipa*
>
> *Well I don't know whether you will receive this letter as I'm going to post it through the civil post at Cape Town or Durban wherever we land for a few hours leave. I hope we do anyhow. I hope I don't have to repeat the same procedure as last time.*

Well it was great to get back into shorts again after 6 months in the cold. We didn't see the sun for the last three weeks we were in England and we were doing outpost duties, which meant sleeping in pill-boxes every fourth night and some nights there was at least three inches of water on the floor so, you can imagine what it was like.

Well I got the biggest surprise I've had for quite a while when I got your cable, Weipa. Thanks old thing, strange as it seems I had forgotten all about my birthday til I got your cable on the 13th, the last day before we left camp.

We have a lot of English troops on with us at present and they are the dirtiest mob of men for soldiers we have ever met and that is saying something. About three days out we struck some rather rough sea and 50% of the Tommies were sea sick. Strange to say I was not. They were sick all over the ship, in the corridors, everywhere so we haven't much time for them.

I have been on A/A for the last week and now I have another week ahead of me so that is alright as it is a good job. God help the plane that tries to attack this boat, there are 16 Brens, 2 Vickers and 2 Lewis machine guns and a 4.5 naval gun on the stern so there will be a sheet of lead all over the boat when they all open up.

It looks like we will be spending Christmas on this 'tub' so that will not be so hot from our point of view. This will be the second one spent in the army. Last Christmas I spent in Bondi and in Sydney. That was before I knew you.

I might see your brother when we get where we are going, unless he is in Greece. Lucky devil if he is. I hope we are there after a few months. We were supposed to be in England in case there were any counter attacks after the Dunkirk stunt and then there were those invasion scares which came very close. At one stage we were called out in the middle of the night and we were standing for nearly three days without a break. I think I told you about that in a letter a long time ago if it wasn't cut out by the censor.

Reading matter at present is very scarce and have had to take to letter writing. I don't suppose it will hurt me though. I should get letters much more frequent than we have been used to lately, we soon will be able to come home for weekends if we get much closer won't we?

Well how is the job going? I suppose it is a bit different from staying home. How is Trudie? Still full of mischief? Give her my kindest regards and I hope she has heard from Jim by now.

At present I have a sore throat and I think it is from too much smoking. I have been smoking a lot lately. I will have to try and knock it off again for a while.

Well I am afraid I have to close this letter now. My brain refuses to think anymore which is nothing unusual for me, is it?

Well so long for the present. I'll send a cable from the first port we land at if I can and in any case I will send one from our destination.

Well so long sweetheart. Remember me to your parents and Beatie.

Lots of Love, Allan

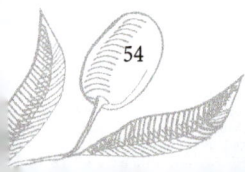

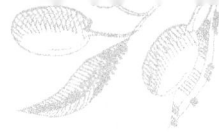

Ronald (Splinter) Collins

I tipped the contents of my bag onto the floor and flopped onto the bed. It was good to be back home in Brisbane after spending the last year studying at Queensland Agricultural High School and College in Gatton.

Looking back on my life, my schooling finished at thirteen and from there I worked a number of jobs. These included twelve months at the Crown Stove Foundry and three years at the City Rubber Co. riding a push bike and carting tyres around on my shoulders. Not long before my eighteenth birthday they sacked me because it was cheaper to employ a junior. Australia was still in the throes of the Great Depression and work in Brisbane was scarce. Life in the country offered more hope of employment, so I got a job as a farm hand on a property at Coulston Lakes. After working in the area until the end of 1938 I decided to return home.

I had enjoyed the work on the farm and talked myself into doing a course at the Gatton School under a government assistance scheme which involved a payment of ten shillings a week. It was there I joined the college-based platoon of the 25th Darling Downs Infantry Battalion at the outbreak of World War II in September 1939.

Standing just over six foot in my socks and being rake thin it didn't take much imagination to see where I gained my nickname of 'Splinter'.

The great unknown

On 3 November 1939, after finishing our exams, a small group of us jumped into my old car and drove up the range to Toowoomba ready to celebrate the end of our studies. Some of us had also decided to take the big step of enlisting. I came away from my college experience with many happy memories, however, nothing could have prepared us for the dramatic change of direction our lives were to take now we were at war with Germany.

Spending a last weekend at home with my parents, sister and six brothers before I reported to the barracks was great. It was hard to imagine how Mum must have felt that day as she looked at us boys playing a game of footy in the back yard. How would she cope if more of the boys decided to enlist?

Fate was to deal her a cruel hand. By war's end Mum and Dad had consented to five of their sons joining the Australian Defence Force and my sister Elsie Jean Collins joined the Land Army. Henry William Collins QX14618 enlisted 5 July 1940; Leonard Charles Collins 404396 enlisted 16 August 1940; Neville Francis Collins QX49899 enlisted 27 February 1943, and Mervyn Thomas Collins 170821 enlisted 16 November 1944.

One day late

Basic training was conducted at Rutherford Camp in New South Wales and then it was off to Ingleburn. Christmas 1939 was spent in Sydney and New Year's Day 1940 in Newcastle. While waiting for embarkation orders to travel overseas about thirty of us ended up in the Prince Henry Hospital for two weeks with the mumps. We were discharged from hospital the day after the troops sailed from Sydney to the Middle East.

Back: Henry & Neville.
Front: Leonard, Ron & Mervyn

The 'mumpy crowd' as we were known, went on to Melbourne and joined the 2/1 Machine Gun Battalion, departing from there in June 1940 for England on the *RMS Ulysses*. The ship stopped in Durban and Cape Town before arriving in Liverpool, England.

We landed right in the middle of an air raid, so we got used to them right from the word go. We were stationed at Bulford on Salisbury Plains before travelling up to Colchester Barracks. This was a welcome change because we got to sleep in proper barracks instead of living in bell tents. I managed to get leave to Scotland with a few mates and visited many of the well-known sights; Falkirk, Allen Monument, Stirling Castle and Edinburgh.

NOTE: The photo above was reconstructed for my parents after the war showing five of their sons in uniform. My brother Henry was taken as a POW by the Japanese on Borneo. He was one of thousands who died of malnutrition at the notorious Sandakan POW Camp in 1945.

Unfortunately our return was hampered by those darn air raids and we were one hour late reporting back. We were sent straight off to the orderly room where I was told to go pack my gear as the next morning I was leaving in the Advance Party for the Middle East.

Embarked on the *RMS Scythia* troopship with 200 other Aussies and 1800 Pommies in a convoy of around 40 ships. Most of the Poms were seasick as soon as we started sailing and not up to eating, so we Aussies ate like kings for the first week or so.

Conditions were cramped and hot, the ship had heaters as it was built for the North Atlantic trade. Having four to a room, roughly 12' x 8', it didn't take long for things to get a bit ripe. We started sleeping on the open deck until a Pommy Ship's Officer (Army) spotted us on the third night and threatened to shoot us if we didn't get below decks.

Whenever an air raid alert was sounded all the ships in the convoy scattered away from us because under power our old tub sent out a long stream of black smoke which could be seen for miles. Once we arrived in Durban it was decided to transfer us to a safer ship, the *Dunera*.

In Durban I met a young lass, Hazel Luck, with whom I corresponded for quite a while. Mum would even write to her but over the years we lost touch. I had only met Hazel the day before we departed so never had much time with her.

We docked at Port Said on Christmas Day and headed onto El Amiriya in the desert. In the desert we discovered that dust storms are worse than sand storms. During a dust storm you couldn't see your hand in front of your face and the dust was so thick. When one was coming you quickly got to your tent and stayed there.

While in the Middle East I was reassigned to the 2/1st Anti-Tank Unit. It was here I met and became good mates with Glen [Stewy] Scott. The next two months were to be like nothing else we experienced in our military careers.

CHAPTER 3

From the Middle East to Greece

The Middle East

(Douglas Channell)

While posted in Jerusalem, Palestine, the A.I.F. were able to conduct a broadcast back home to Australia. The program in question was called 'Voices from Overseas'. Nurses and men of the A.I.F. who were stationed on the front in the Middle East were given the opportunity to record messages. These messages were then broadcast back to Australia via the BBC. It was extremely gratifying for their relatives and friends to be able to hear the voices of their loved ones from so far away. This was an unexpected lifeline for families waiting for news.

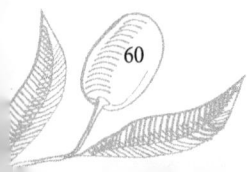

Doug in Cairo

"Channell, you've had experience working in radio, would you like to organise and compere the upcoming broadcast?" I was happy to comply and although no names were mentioned, my colleagues back home knew exactly who was hosting the program. It was a very lively performance with a number of our men playing musical instruments and singing songs from home including numerous impromptu and highly raucous numbers we had composed ourselves. Others told stories, recited poetry or did impersonations. All up it was a thoroughly uplifting event for both those at home and us here on the war front.

... Band Broadcasts. The A.I.F. Band broadcast from Jerusalem under the direction of Lieut. Douglas Channell, well known in Sydney. The recital began with a fanfare of cooees, then played Gundagai, Waltzing Matilda, and a number called The Vanishing Army – referring to the A.I.F. "somewhere in Palestine". Private Connolly, a dairy farmer from Penrith, gave a spirited imitation of the kookaburra. There were messages by Brigadier Allen and a battalion colonel who served in Palestine in the last war. Lieut. Channell concluded the broadcast by saying, "We send our regards to Hitler, Goebbels, Himmler, Hess and the rest of the mob. We hope our next broadcast will be from Hamburg, where we'll take the place of 'Lord Haw-Haw'."

"Jobs At Home For Men Of 2nd A.I.F." *The Courier-Mail*

I interviewed members of the band and asked them to discuss what they had been doing before they enlisted and their reasons for enlisting. No names were mentioned but I recall some of their comments, which went along these general lines:

"I wanted to do my bit and see the world on five bob a day..."

"I played cricket for Australia the week before I enlisted..."

"I fought here in Palestine in WWI, you wouldn't recognise it as the same place..."

"My father and uncle fought in the last war, now it was my turn..."

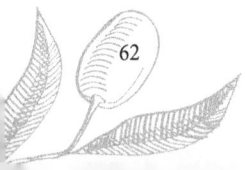

Lord Haw-Haw

I believe at this point I should explain who I was referring to when I spoke about Lord Haw-Haw. Lord Haw-Haw was the alias of American born William Joyce. Born in Brooklyn in 1906, Joyce and his parents returned to their native Ireland in 1909 before eventually settling in England. Joyce worked as a radio announcer and tried his hand unsuccessfully as a politician. He embraced the ideology of fascism and had a reputation of being a bully and a fighter. During a fight at a political rally someone slashed him across the face with a knife, scarring him for life. Once Germany announced it was going to war in 1939, he and his wife left Britain and travelled to Berlin.

After his arrival he very quickly secured a contract as a newsreader at the *Reichsrundfunk* (German Radio Corporation). Here he worked for the Nazis spreading propaganda. He conducted a regular radio broadcast into England under the banner of 'Jairmany calling'. It was here he sprouted his propaganda urging the British people to surrender.

He would proclaim earnestly how Germany and the Third Reich were superior and describe how they would destroy England and the Commonwealth. His tone was menacing, jeering and full of sarcasm. Initially the English population listened to the broadcasts and considered them entertaining, but as the war continued they ceased to see the funny side of things and stopped listening. This was never more evident than when Germany started bombing England. Though Lord Haw-Haw continued to broadcast into England from Germany, his propaganda for the most part fell on deaf ears. Everyone now knew in no uncertain terms what a traitor he was to the land of his birth.

ANZAC Corps

Britain was now in a position to honour a pledge to its only ally on the Continent, Greece. This was achieved by sending troops to support Greece from the planned invasion of Germany and Italy. Initially Britain sent a small forward party of non-combatant personnel to Greece to gather and access vital information for the Empire.

On 7 March 1941, Prime Minister Winston Churchill diverted 58,000 Allied troops from Egypt sending them to occupy the Olympus-Vermion line. The operation was code named Operation Lustre and included 17,000 Australian and 16,700 New Zealander servicemen and women. The Australian 6th Division and the 5th New Zealand Brigade Group now fought as an ANZAC Corps under a unified command, the first time since Gallipoli.

Over the next two months the following soldiers were for the most part to be involved in the same conflicts.

~ Andrew Findlay Borrie, Motor truck driver New Zealand Expeditionary Force (Greece, Crete, POW)

~ John Borrie MBE New Zealand Medical Corps (Greece, POW)

~ Alfred Clive Carpenter, 2/4th Australian Infantry Battalion (Greece, Crete)

~ Douglas Ronald Channell MC 2/1st Infantry Battalion (Greece, Crete, POW)

~ Ronald (Splinter) Alexander Collins 2/1st Australia Anti-Tank Regiment (Greece, Crete)

~ Alan Eason 1 Corps troop supply, Australian Army Supply Column (Greece, POW)

~ Albert William Mayer MID 2/4th Australian Infantry Battalion (Greece, Crete)

~ Clifford Harrie Morris 2/5th Australian Infantry Battalion (Greece, Crete, POW)

~ Reginald Walter Saunders 2/7th Infantry Battalion (Greece, Crete)

~ Glen (Stewy) Derek Scott 2/1st Australia Anti-Tank Regiment (Greece, Crete)

~ Dimitri James Zampelis 2/2nd Field Regiment (Greece, Crete).

Land of the Greek Gods

(Glen Scott)

Splinter and I saw no fighting in Egypt but did some recovery work for field workshops. Little did I know the wish I expressed to Weipa to join her brother, Albert (Gus) William Mayer (my future brother-in-law) of the 2/4th in Greece was about to become a reality. When news filtered through to the lower ranks that we were heading to Greece, we all shouted for joy. Heading to the land of the Greek Gods, what more could any red-blooded Aussies want. We left for Greece on 1 April 1941.

It was exciting to finally be on board a ship heading towards the rugged coastline of Greece. As was always the case we were briefed about the people, cultural differences and the climate. No sand stretching out endlessly and none of the dust storms we had experienced in the Middle East.

Headquarters relayed a special briefing from Lieutenant-General Sir Thomas Blamey, Commanding Australian Imperial Force (ME) on 4 April 1941. Briefings can sometimes seem like a waste of time, however it would soon be clear to each and every one of us just how true these words were.

His directive read:

> "Just twenty-six years ago the Australian Army carried out its first great operation on northern Mediterranean shores when our kinsmen of the 1st Australian Imperial Force landed on Gallipoli. We have now landed again in these regions to fight alongside the Greek Army to overthrow once more a German effort to enslave the world.
>
> The Greek nation, the smallest and poorest of all the nations that the Axis powers sought to bully into submission, alone in southern Europe, refused to submit.
>
> Their efforts, along with our own in Libya and Abyssinia, backed by the valour of our Fleet, have already destroyed one of the Axis partners as a power and have forced the Germans to take control of Italy's destiny.
>
> There can be no doubt also that their valiant and successful struggle has had a great effect in determining Yugo-Slavia, after having yielded to German bullying, to arise and defy the Axis powers.
>
> In Australia we had a very wrong impression of this valiant nation. I am sure that, as you get to know the Greeks, the magnificent courage of their resistance will impress you more and more. It is not unlikely that the action of this small but noble nation may prove in the end to be the beginning of the final downfall of Nazi tyranny.
>
> Before you are long in Greece you will realise that every Greek man and woman and every pound of Greek money is being put into the effort to win the war and that they are undergoing great privations and willingly making great sacrifices to do so.

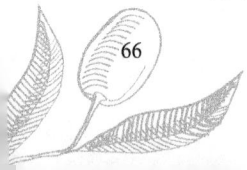

I am sure that this will lead every Australian, worthy of his race and country, to regard every Greek man and woman with friendly eyes and to treat their institution, customs, and manners with respect. We come to them as deliverers and they welcome us as such.

Let each of us, therefore, so conduct himself as to ensure that we shall hold their respect and friendly goodwill as long as we shall remain in their country, that we may fight side by side with complete confidence in one another."

TA Blamey

Source: Documents supplied by Roxane Scott

6th Divvie on the move

On the day of our approach to Piraeus, the sea was calm and flat. Fortunately, there were no enemy aircraft around as the ship manoeuvred safely into the harbour.

We landed at Piraeus on 2 April 1941 and were welcomed by the Greeks with open arms. We disembarked and set our feet firmly on Greek soil. Greece was the home of so many ancient monuments and stories, myths and legends we had grown up with. It was a surreal experience to be standing in Greece amongst it all.

On the foreshore there were many large stone buildings, with hills rising up high behind the houses. Such a vast difference to the landscape and timber houses back home in Australia. Greece is very mountainous and rocky in comparison to our homeland of predominately wide open spaces.

As we drove through the streets in convoy to our camp the women and children threw flowers at us and the old men cheered us on. It was evident most of the men from the region were off fighting.

The genuine friendship and goodwill shown by the Greeks only strengthened our resolve to forge strong ties with these people and give our all in the coming encounter with the Germans.

After we were moved by military transport to the bivouac area at Glyphada we had a little time to ourselves. This was to be our only time here and while we accomplished a little sightseeing during our two-day stopover, we made a point of visiting the city bars and cafes. Soon these businesses were all crowded to overflowing with allied soldiers. Greek civilians and soldiers joined together to share their strong local wine called Retsina. We found it quite rough and some commented it tasted like paint stripper and boy did it pack a fair punch. There were a few sore heads the next morning.

Greece is a country not only with a rich history but also one full of soul. As soon as we landed at Piraeus we were greeted with the 'thumbs up' sign. One of the forward non-combatant troops had taught them by using this signal it didn't matter that language was a barrier, the 'thumbs up' sign would act as a symbol of friendship.

Though the conversations were broken and limited because of the language barrier, the messages from one to the other were understood. We all enjoyed ourselves during this small rest stop on our journey to the front line. It was the first time I experienced the taste of olives and feta cheese. The food was plentiful and quite delicious though a bit strange to our tastes. It was a welcome break from the same old army rations.

In so many ways these people made us feel at home. They were so grateful to see us, recognising we had come to join their husbands, fathers, brothers and uncles to fight their enemies and allow them to keep their freedom.

The 2/1st Anti-Tank Regiment head north

Our objective now was to journey to the Vevi Pass on the northern border of Greece and Yugoslavia, where preparations for Operation Lustre would be put into action. Hopefully the Yugoslavian soldiers would be able to hold off the German attack, at least for the interim.

As troops from the 2/1st Infantry (Douglas Channel), 2/2nd Field Regiment (James Zampelis), 2/4th Infantry (Alf Carpenter and Gus Mayer), 2/7th Infantry (Reg Saunders), 1st Supply Corp (Alan Eason) and the New Zealand contingents (John and Andrew Borrie) arrived and joined the 2/1 Anti-Tank Unit (Splinter & I) we either entrained or drove transports north to the Vevi Pass.

The 2/1st Anti-Tank Unit drove our transports along and it seemed to me as I rode along on my motorbike that fruit trees had been planted wherever possible. Our convoys would often pass mountainsides covered with olive, lemon and fig trees in amongst the bushes and native vegetation. Little did we realise at this time how our very safety would soon rest with those olive trees in the coming weeks.

Stewy beside his bike

In most of the villages we passed through, people were going about their everyday chores. They could be seen leading donkeys carrying heavy loads of produce; women sat spinning yarn from wool shorn from their goats; old men, younger women and boys were working the fields; old women could be seen picking herbs for cooking while others sat in the shade sewing.

Vines and creepers grew on wooden frames outside the houses. This was where they gathered in the early evening to share a drink after a hard day's work. Outside the houses small children played on the ground with pebbles. A simple but rewarding life was led by these people who lived to work and keep their families together. They all waved and cheered us on as we passed by.

In the proximity of Larissa, the capital of the Thessaly region, I experienced my first earthquake. I can tell you it was not just a tremor, as I was thrown from my motorbike.

Conditions were not easy for the transports as they had to contend with the rough mountainous roads. Continuous cold weather including ice and snow was not something we enjoyed. Units using the railway fared a little better than those of us transporting guns and equipment.

The nearer we got to the border the more mountainous the terrain became. The going was difficult as the primitive roads we traversed deteriorated quickly, as they were not built to carry heavy motor-transports. Driving without lights along ridges at night, through deep valleys and ravines was hair raising at times but we were forced to do this in order to keep our positions hidden from any enemy planes. Going ever forward was necessary if our units were to join up with the Greek soldiers and other members of the 6th Division already stationed at the northern border of Greece.

Keeping the movement of heavy guns proceeding during the night was difficult enough without people, horses, mules and cattle all trying to use the road as an escape route at the same time. The closer we got to the border the more the traffic coming from the opposite direction increased. It consisted mainly of refugees from Yugoslavia. Their military personnel insisted on 'right of way' until convinced otherwise by us. The refugees also made night travel fairly hairy as those in vehicles insisted on driving with their headlights full on. This error was corrected forcibly for the safety of everyone.

It was an appalling sight, watching the endless stream of homeless women and children trudging the roads in the bitter European winter, desperately trying to flee the invading German army. Little children, tired and hungry, begged not to be left behind. It was a sight we knew we would carry with us to our graves.

Eventually we arrived at the border village at the Vevi Pass. The next couple of days saw us digging in and setting up our fortifications. Once this was completed we joined forces with the engineers and together we started to bury mines on the road across the border. There was no time to improve the roads so the 2lb anti-tank guns had to be hauled into position for backup. These guns were placed so we could cover the railway and the road through the Vevi Pass. Our unit was kept busy making sure the trajectory, height, angles and bearings of the guns were set correctly in readiness for the coming assault. This position was further fortified and consolidated by a British Sherwood Foresters Infantry Regiment and a New Zealand Vickers Machine gun platoon.

2/4th Infantry Battalion

(Alfred Carpenter)

Gus Mayer and I left Alexandria on the SS Pennland and arrived in Greece on 3 April as part of the 1,100 strong troop of the 2/4th Australian Infantry Battalion. Looking around I commented to Gus on the beauty of the landscape in this part of Greece. We had two days leave and set out to enjoy every minute of it.

Upon arrival I received word that I was to be promoted to Regimental Sergeant Major. I bumped into and shared the news with an old mate of mine, Max, who like me, hailed from Wagga Wagga. It turned out Max was commandant at a nearby monastery where the monks were self-supporting. They had their own cattle, chickens, vegetables and grapes. Better yet they made their own wine and we were very happy to sample it. A gathering to celebrate my promotion was swiftly put into action and thankfully the offer of a bed for the night was thrown in. I wasn't real well the next day, I can assure you.

Group photo of 2/4th

With the celebrations behind me, it was my duty on 5 April to get the men from the 2/4th entrained so we could set off for Larissa. Conditions were a little cramped on the cattle train but as it rumbled off into the sunset, it was possible to see fields planted with crops, wildflowers and the snow-capped mountains in the distance. At this stage, Greece was not yet at war with Germany.

Gus and the members of the 2/4th Transport Division had earlier headed off by road and were set to join us at Larissa. Once we all arrived we continued on and joined forces with British, Dominion, New Zealand and fellow Australian soldiers with artillery, tanks, our engineers and combined infantry. Our orders were to delay the advance of the Germans for as long as possible at the Vevi Pass.

By now the Australian units spread across the hills at Vevi included the 2/1st Anti-Tank Regiment, 2/1st Infantry Battalion, 2/2nd Field Regiment, 2/3rd Anti-Aircraft Regiment, 2/4th Infantry Battalion, 2/5th Infantry Battalion, 2/8th Battalion,1 Corps. Troop Supply, as well as New Zealand units, British Sherwood Foresters and the Greek Dodecanese Regiment. Allied air support in Greece was minimal so the defence of Greece, for the most part, would be carried out by ground combat troops.

Under the command of Major General Ivan Mackay, preparations forged ahead at Vevi. Gus was part of the contingent setting up gun placements to cover the railway, while the engineers, tanks and infantry units were all working furiously to get everything in order. Things were heating up. It wouldn't be long now before we saw action.

Digging in

(Glen Scott)

Conditions near the border were brutal. Because our positions in the Alps were, on average, around three thousand feet above sea level, we all struggled with the rarefied air. Lack of oxygen caused many problems as it took some getting used to. Heavy snowstorms compounded the situation and didn't make the job of digging trenches or paths any easier as the ground was as hard as concrete.

Splinter commented it was bad enough trying to concentrate on watching out for the advancing enemy, when all the while you were standing in the trenches up to your knees in icy slush with either snow or rain falling on you. Trench rot and frost bite were painful problems for many of the men and the cold winds were relentless; not something we Aussies were very familiar with. The only way to keep warm at night was to double up. Two men slept side by side sharing two blankets; it was better than freezing.

When snow fell on the hills during the night, we awoke to find inches of snow lying on our greatcoats and blankets that covered us. Water in our bottles would be frozen unless you had thought to place it under the blankets with you.

I was luckier for the most part as my main job was as dispatch rider for our unit, requiring me to ride between units with messages. Don't get me wrong, this had its own dangers and challenges, but it was better than constantly having to stand in freezing mud and slush beside the anti-tank guns like Splinter and the others. The freedom of the bike also allowed me to pick up some rations here and there, which I would share with Splinter when I returned to camp.

None of us knew how the Greek soldiers had coped with the savage blizzards for so long. It was a pitiful sight to see their packhorses standing one minute and the next the poor beasts would lie down and freeze where they fell. Hunger, cold, exhaustion and lack of rations was getting to be more of a problem each day, but we knew we had a job to do and do it we would.

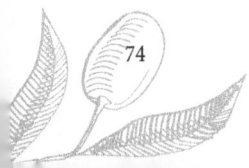

Snippet of news from Greece

In a letter received at the ABC in Australia, Lieutenant Douglas Channell, 2/1st Infantry Battalion wrote about his unit's arrival at Vevi Pass.

> *'Bill Travers' company and mine did a forced march over mountain stream and a river – 34 miles in 12 hours to join the rest of our battalion, and to take up another position in the snow to stop the advance. Very proud of that effort are we, for we were in full battle order and carried rations as well as a blanket.'*

"THE MAN IN THE STREET" *Smith's Weekly*

Germany invades Greece

Yugoslavia had little hope of holding the Germans back from the Greek border as their soldiers were relatively few in number and ill equipped compared to the might of the German army. The Greeks concentrated on defending the north-west border between the two countries, which left them extremely vulnerable as they were spread very thinly on the ground. It did not take long for the organised, well-equipped German panzer divisions to break through their defences. On 6 April Germany invaded Greece.

Last ship from Egypt

(Reg Saunders)

Back in Athens the last of the troops were arriving from Egypt. I was a member of the 2/7th Battalion under the command of Colonel Walker.

I had a lot to be thankful for, having just spent time with my younger brother Harry (Harry Saunders VX18629). Harry had been in a rush to enlist, convincing the recruiting officer he was old enough and joined the 2/14th Battalion. Harry and I met up while his unit was camped near mine in Palestine. We were able to take leave together and spent time in Jerusalem touring the sights.

When I returned from leave in early April my unit was being deployed to join in the Greek offensive. On the day our ship arrived it was impossible to dock at the wharf because of the number of ships already in port. This necessitated us having to disembark one boat at a time, which was a very slow process. It was no fun at all, particularly for those of us who ended up being left on board overnight. We were ordered to carry all of our gear down to one central place inside the hull. I remember it being very hot and airless below deck, a stark contrast to the cooler conditions experienced top-side.

The nurses aboard with us had been taken ashore in some of the very first boats. These women were a top group and they willingly went wherever they were directed in order to care for our sick and wounded. Some of the conditions they were forced to work under were at times primitive to say the least. You never heard them complain; even when they were caught up in extremely dangerous situations in the Middle East they had soldiered on doing their jobs to the very best of their abilities. It was to be no different here, as Germany had only days before declared war on Greece.

From the first day we were under constant attack. Not knowing what lay ahead of us, we were relieved to know we had such good medical support. We would be relying on their skills and knowledge in the weeks ahead.

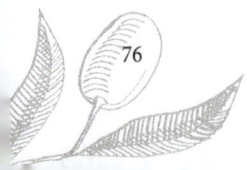

After spending a very uncomfortable night on board; the following morning after breakfast, it was our turn to disembark in the pouring rain. Once assembled on shore we marched a short distance to trucks that transported us to the camp.

How different the landscape was from Egypt. Athens was much larger than I thought it would be and a beautiful place despite the rain. Once in camp we were allocated to our tents and at the end of a long day spent little time bunking down for a good night's sleep. We awoke in the early morning to the sound of an air raid. The Germans were bombing and strafing the camp. What a welcome to Greece!

Thankfully the raid didn't last too long and some of us managed to catch a lift into town. Everyone was so friendly and warm in their welcome and generous with their time. Even though language was a hindrance on both sides, we were soon able to establish we were after something to eat and drink.

Early the following morning we marched out to the railway ready to head north, but air raids with over twenty planes on the attack delayed our departure until later in the afternoon.

It was a great relief once we finally got under way. We were on the train for the next few days; it was a long slow trip but the countryside was beautiful. Because of the air raids during daylight hours we often had to stop and run and hide in the countryside in the pouring rain. It was impossible to get dry under these conditions and one just had to grin, bear it and get back on the train, inching closer and closer to our destination, the Vevi Pass.

Make a Dedication

Troops of the 2/7th Infantry Battalion in the carriages and on the station platform waiting for the south-bound leave train to start. Shown: NX111740 Private (Pte) R. Chapman (1); Corporal (Cpl) Howsan (2); VX36219 Sergeant (Sgt) A. S. Haynes (3); VX13068 Sgt Congress (4); VX12843 Sgt Reg Saunders (5); VX43127 Pte L. Brodie (6); VX12370 Cpl. J. W. Jones (7); VX59057 Pte A. M. L. McLuckie (8); VX13672 Pte R. Parsons (9); VX55851 Pte Farnell (10); VX556791 Pte C. J. Spence (11); VX44293 Pte R. Wilton (12); VX33134 Pte A. G. Dorecott (13); NX58704 Pte N. Watson (14); Pte Toynton (15); VX16560 Pte W. Kann (16).
Australian War Memorial - Accession Number 057894 Maker Tait, James Place made Australia: Queensland, Australia: Queensland, North Queensland, Innisfail Date made 12 October 1943

Reg and group of men at train

On our way north we came across large numbers of our non-combatant troops heading south. At our first designated stop it quickly became obvious we were in deep trouble. Air raids of anywhere up to thirty planes, bombing and strafing were consistent for two to three hours at a time. The dive-bombers were the worst, you could hear them long before you saw them and it sounded as if they were heading straight for you. A terrifying sound and nothing you tried could get that sound to go away. The 2/7th were in the thick of it now!

Action on the border

(Glen Scott)

The Greeks put up a spirited defence on the Metaxas Line. They were housed in cleverly camouflaged positions commanding views of main entry points into their country. Despite being attacked by heavy artillery detachments they doggedly held on. In some cases the Germans had to continuously bombard the forts for over thirty hours in order to take these positions. The Greeks put up a bitter resistance and losses to both sides were heavy.

Within days all Allied units at Vevi experienced their first taste of full on German fire-power. There was limited air support from our side and on one occasion, one of our Spitfires was brought down by enemy fire as we watched on in despair. Unluckily for the pilot, he crashed landed in our own minefield with spectacular results. A recovery party from the 2/1st accompanied by a medical orderly quickly went into action and was able to bring the pilot safely back into our lines before he could be captured by the Germans. By chance and good fortune he had managed to escape serious injury. Distressingly we lost many of our brave pilots and crews during this encounter. The fire-power thrown against them was immense and our planes were ruthlessly cut down. They stood little chance against the superior number of German planes.

Splinter was standing by his anti-aircraft gun ready to engage with the Germans. I was overlooking the valley and could clearly see the German vehicles advancing south. I realised from the direction the Germans were coming, that despite camouflage covering a couple of the guns the 2/1st position would be painfully obvious for the enemy to observe. After speaking with the OC it was decided to reposition these guns and substitute dummy guns in their place. This deception had the desired effect and the dummy guns were subsequently heavily engaged by enemy artillery and mortars totally destroying them. As the Germans were firing at the dummy guns, we were able to engage our well-camouflaged guns to fire on the German planes. We were successful in inflicting some worthwhile direct hits and bringing down a number of their planes.

Next morning, a German half-track troop carrier ran into our minefield. It was overturned and blocked the road. Our 2lb anti-tank guns engaged another tank behind it. The Germans shelled the rest of the minefield with instantaneous fused shells and made the use of a small protection squad inoperable.

Disappointingly the little air support we had was unable to operate over the next two days due to low cloud conditions and heavy falls of snow. We had thought the cold and snow was bad before but now it was bitterly cold and the strong icy winds were almost intolerable.

These conditions worked in favour of the Germans, allowing their Alpine troops the opportunity to actively patrol followed by heavy bombardment with mortar fire on our positions. They were tricky buggers. They would call out in perfect English, making us think they were some of our men. Initially if they received a response, the Alpine troops were able to infiltrate our positions from behind and start attacking us with small arms fire.

Other times they dressed as Greeks and tried to gain access behind our lines where they would then turn around and start shooting. We quickly woke up to their tactics and maintained a code of silence.

When the German forces started to break through the Allied defence, the 2/4th with Alf and Gus moved further across from their positions in an effort to maintain ranks. A small force of RAF fighters and bombers supported our ground troops where they could. They gave the Nazi soldiers a taste of what we had been receiving. Let them see how they liked it when the shoe was on the other foot.

When we opened fire with the artillery on the morning of 10 April, the company commanders on the ridge east of our positions could see the 2/8th wearily scrambling up the hills to fill the gap on the right of the lines. After an eleven-mile march they still had a steep climb to reach and dig in to the positions allocated to them. Not an ideal situation for them by any means.

The Germans were throwing everything at us now: spotter planes, bombers, hundreds of tanks and thousands of soldiers continued advancing once they were through the pass. Overall the general situation was rapidly deteriorating and after three weeks of hard constant fighting with little success in maintaining and defending our positions, withdrawal was deemed necessary.

CHAPTER 4

Withdrawal from Greece

Leaving Vevi

(Glen Scott)

Greek forces were under extensive attack. Every one of them fought valiantly to defend his country, in many cases resorting to hand-to-hand combat. These Greeks, who for the most part were fighting with pre-World War One rifles, were no match against the sheer number of Germans with their modern equipment.

Outnumbered 100 to 1, the Allies didn't stand a chance. Splinter and I both realised stopping the Germans here was now a lost cause. The worsening situation marked the start of a disciplined withdrawal of our troops from central Greece.

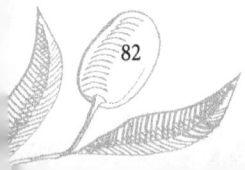

One incident in particular showed just how on edge our troops were. I received instructions to hop on my motorbike and deliver a message to the person in charge of the gun tractors advising them to take the tractors up to the front line in order to get the guns out. Having parked my motorbike, I was walking to a gun position with the message to prepare to withdraw when I was taken prisoner by some English field police as a suspected spy. I thought the situation was ludicrous until I heard the hammer of a .45 click and felt the gun in my back. Fortunately for me, the gun section sergeant knew me and came across to explain I was Allan Stuart from the 2/1st Anti-Tank Regiment. This all happened because I was wearing an issue driver's cap instead of a tin hat. Nervous times indeed!

While riding back to the transport lines I was blown from my bike by a 25lb battery strength fire blast. On recovery, I found the bike was useless, so I walked all the way back to the transport lines where I was issued with yet another replacement bike.

We realised too late the position where the transports had been parked was flat and open. Almost half of the transports became bogged when the drivers tried to get them to the road. Because of all the bombings the roads were a mess. While trying to extract these vehicles they came under artillery fire destroying many of our transports. The 2/1st Anti-Tank Regiment went into action with forty-eight guns and would leave with fewer than half the number.

During the evacuation we were pursued by Stuka dive bombers. They fired on us continuously until the evacuation from the front was achieved. We also encountered high level bombing by Dorniers and Heinkels and low level strafing by Messerschmitts, which could not be mistaken as they all had yellow painted noses. I never discovered the reason for the yellow noses.

The previous night's hot meal was to be our last until we landed back in Egypt. Food was desperately short and a tin of bully beef could be traded for almost anything. Our staple combat rations were bully beef and hard biscuits. Standard intake was one tin of bully beef and one packet of biscuits between three men per day. There was to be no M and M stew as it was impossible for the field kitchens to operate during the withdrawal.

The Germans were attacking us relentlessly from the air while their troops, tanks and heavy artillery moved quickly through the mountain pass. Our orders to slow them down as much as possible were getting harder and harder by the hour. It was a terrifying experience, difficult to comprehend. It was sheer courage that kept us going, while having to acknowledge defeat, especially coming off the back off our success in the Middle East.

The Allied Forces were able to hold off the German 6th Mountain Division for three days. Our strategy of short bursts of fire and defensive actions was followed by strategic withdrawals through the valleys and down the mountains. This came at a great cost for the allies, with the heaviest losses of life and wounded troops in Greece.

Our engineers buried as many land mines as possible and as we drove away it was not long before we heard the sound of mines exploding. Once the Germans ran into a mine field they were forced to take time to destroy the mines before they could continue their advance. Every little bit of sabotage we were able to initiate helped with our extraction.

Every unit experienced the same problems during the withdrawal. The methodical management of this enormous task involved forced marches through snow clad mountain passes. This sometimes required soldiers to walk distances of up to 34 miles (55 km) at a time before they were safely behind Allied lines. It was a very dangerous and exhausting experience. As soldiers we were determined to show the invading Germans we were equal, more than equal to slow their greater numbers of better-equipped soldiers.

Full retreat

The first leg of the retreat from around the Vevi Pass for the 2/1st Anti-Tank regiment was through the town of Florina 13 miles (22 km) to the west. Along the way we discovered all the food for the division in dumps, which the engineers were preparing to blow up. All trucks going past were invited to load up with as much as possible before they destroyed what was left so the Germans wouldn't be able to utilise any of it. There were foods Splinter and I had not seen since arriving in Greece, including tinned fruit and cream. We also found some clean clothing, allowing some of us to change out of our filthy uniforms.

The majority of our travel had to be accomplished at night owing to the daylight air attacks. As the battery scattered after each confrontation, my job was to jump on my motorbike and get them regrouped, which proved to be quite some job I can tell you. Twenty-four hours without sleep was not unusual and at times, the weather conditions made life more than a little uncomfortable.

On the move

(Ronald Collins aka Splinter)

"Collins get that vehicle moving, now!" I was ordered to continually position my vehicle so the other gunners working with me could hook it up so I could tow the anti-tank gun during the retreat. When it was our turn to cover the withdrawal from the rear I would have to set up the gun and cover the retreat of those following us. Once they were safely passed it would then be the turn of the unit behind us to cover our retreat. Pack up, hook up the gun and drive off once again. I reckon it was one of the most frightening experiences I have lived through, trying to tow the gun and avoid the German artillery fire coming from the rear.

I found the constant night driving with no lights exhausting. Roads had been smashed to pieces because of the German bombardment. It was hard work concentrating on moving forward when all around me I could hear the horrible screams of the bombs, not knowing where they would land. Would my truck and those with me be the next to be killed?

Ordinarily if one of us lit up a smoke, you'd pass the match around so three people could light their cigarettes from the one match. Most of us smoked as it helped to calm the nerves. Nevertheless one of our snipers advised me to only light two smokes from the one match at night time so any German snipers wouldn't have enough time to draw a bead and shoot the third man. Smoking at night in the open was not encouraged.

Trucks not lucky enough to escape the bombing marred the roads and sometimes I was forced to stop so we could bury our dead. I would record their names and co-ordinates of the burial site and later hand these records on to the CO. Mangled remains of trucks were shoved to the sides of the road to allow our trucks through. Emotions ran high at times like these, but we had a job to do and the Germans were not going to stop us from carrying out our orders. The longer we could hold them back the more chance our troops had of withdrawing and by goodness we would do our job to the very best of our ability.

Larissa to Lamia

(Glen Scott)

During the four days before the 2/1st arrived at Larissa, it had been observed by the advanced platoon, that while the town had been subjected to heavy levels of bombing attacks, only the hangars, aircraft and the RAF camp at the airfield were affected as the airfield strip itself was not damaged in any way. Having been present when the intelligence report about the airstrip was read out, I knew it to be accurate.

I rode my motorbike into Larissa ahead of the others and found the next food dump. The Australian Canteen Depot was full of beer, tobacco and cigarettes, all there for the taking. The men were pleased I had found the depot and cheered, "Thank goodness for Stewy" as they loaded up with as many provisions as they could carry.

Many things happened at Larissa that should not have happened. For instance I was directed to the wrong road by the MPs and was almost back in Yugoslavia by daylight. Realising what had happened and while turning around I almost ran into a German motor patrol, luckily I saw them first and was able to hide.

Between Larissa and Lamia, our unit met up with a number of other units including those of Douglas Channell, Alf Carpenter, Gus Mayer and Jim Zampelis.

Captain Douglas Channell and the men of the 2/1st Australian Infantry Battalion under his command were on the move and retreating from a slightly different direction to my unit. Captain Channell described the situation as:

> *"...being impossible to travel much in daylight hours because of the constant barrage of strafing from the German airplanes. The men were getting hardened to the trail now and felt more at ease travelling in the dark. Every time we crossed one ridge we could tick it off as one less hurdle to conquer. The pace quickened in the pre-dawn, as we clambered down timbered slopes, over broken, jagged rock-strewn country absolutely destitute of human inhabitants. We were all fleeing the Germans now and had the same goal in mind – get to the bottom of Greece to the ships waiting to evacuate us to a safer place."*

> *Source: Channell family papers*

The air raids became heavier and many casualties were sustained from the dive bombing and strafing. This was especially cruel when our transports were caught out on an open road or in a cutting. The poor devils never stood a chance!

Somewhere near Lamia an aerodrome with Glen Martin fighter bombers all intact and fully fuelled and serviced with ammo was discovered. These fighters were burnt and the engineers blew up some of the bomb dumps, leaving nothing of use for the Germans.

While we were stopped here, the 2/1st shot down a German Recci plane, killing both pilot and observer. Within half an hour we paid for it dearly with dive bombers, strafing and high level bombing, losing more transports, men and stores, plus my third motorbike, which had the tank and head blown off. A piece of shrapnel lodged in my leg as I was flung into the air. I pulled the shrapnel out and applied a field dressing. The next day I found another bike and continued my duties. This was more important than ever as the telephone lines between Battalion HQ and Company HQ were by now out of action. Men on bikes or foot soldiers were now the only method of communication between units.

Splinter and I were able to breathe a sigh of relief once we finally reached the embarkation point at Megara. Unfortunately because of space limitations on board the ships sent to rescue us, we were ordered to leave behind what remained of the transport and guns we had taken so much trouble to haul with us. We couldn't do much about destroying them as we were hidden in olive groves to give us partial cover from air attacks. What we did do was disable every item we had to leave, making the transports and guns useless and inoperable. Sheer exhaustion from the constant slog came close to overcoming us. Our limited rations, firearms, blankets and great coats were the only things we could take on board, I couldn't even take my motorbike.

The anti-aircraft ship *HMS Conventry* and other convoy vessels arrived in the middle of the night. In order to board the ship it was necessary for us to wade through water up to our necks, before clambering up rope ladders slung over the side. Even carrying as little as we did, this used up the last of our strength and we lay on the crowded decks where we could. The ship sailed for Crete on 25 April 1941. An ANZAC Day we would never forget.

The naval crews did a wonderful job getting us to Crete safely. They then turned around and headed back to Megara to collect others waiting to be rescued. Before leaving Greece with yet another shipload of soldiers they were able to shell the area where we had been forced to leave our trucks and equipment.

Following is an excerpt from the poem, *The Sixth Division Saga – 1939 to 1945* written by QX4567 Ex. Sgt. T.J. Kemp 2/1/st Tank Attack Regt 6 Div. A.I.F.

… To Greece our orders said was next, to stay the mighty Hun,
The gallant Sixth across the Med. took the badge of the rising sun.
With odds of nearly ten to one we met them mid the snow,
The ANZACS, British and the Greeks fought battles we all know.

For fourteen days the German hordes came at us day and night,
The Luftwaffe fighters gave us hell, their bombers showed their might.
There simply was no panic; we held up rather well,
We fought when we were needed in those youthful days of hell.

Withdrawal to the south, a deadly game of chess became,
"Blow the bridges, slow the Germans, save light weapons all the same!"
Our mighty Navy waiting, to their nets we wade and climb,
We hardly had the time to think of mates we left behind.

Our transfer from destroyer to a troopship made at dawn,
Saw the Stukas dive to aim their bombs to scream upon the morn.
Our escort warship's ack ack guns, our Brens and rifles too,
Made them change to pattern bombing they were losing quite a few.

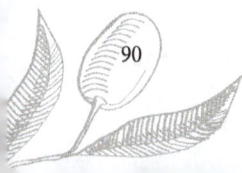

The walking wounded

(Jim Zampelis)

As the balance of the 2/2nd escaped south, I did whatever was necessary to help soldiers along the way. While the trucks continued to slowly head south, I would jump out of the truck and help weary, battle-fatigued soldiers walking along the road to get up into the trucks. As the regiment's cook, it was not my duty to do this, but as I couldn't cook I needed to do something. Later my commanding officer addressed me in conversation saying he was impressed by the calibre of my mettle. He told me how the regiment's medical officer had reported to him how impressed he was with what he considered a great undertaking of outstanding bravery in the face of extreme danger on my part. My unselfish actions and fighting spirit had saved many lives and for this he was thankful. I was only doing what I believed was the right thing by our troops.

Eventually overwhelmed by the German forces, a large part of our unit crumbled and many of our soldiers were left trapped behind enemy lines. I was told fourteen of the men from my unit had been killed in the confrontation and over 100 were taken prisoner. Thankfully I was amongst those sent forward, as we were mainly non-combatant troops who were sent forward and escaped that particular conflict.

We escaped the Germans and travelling miles out of the way, reached the beaches and were able to join others at the assigned spot. From here we evacuated.

The Greeks helped many of our troops who had been cut off from their units during the withdrawal process to hide, escape detection and get safely to evacuation spots. Being of Greek descent myself it was particularly heartening. I for one was very grateful for their help and thanked them from the bottom of my heart.

Aliakmon River

(Alf Carpenter)

Meanwhile further north, the 2/4th continued to carry out critical rear-guard actions. As RSM of the unit, I noted in my diary how we had been overrun by the Nazi's tanks and motorised infantry. I, and a quarter of the men, managed to escape during a fighting withdrawal and formed up close to the Aliakmon River. The officer in charge behind us, evaluated the situation and with little hope of escape from the Germans, and in order to save the remaining men, he surrendered.

Looking at the conflict through my binoculars, I witnessed the Germans line our blokes up and use them as human shields to protect themselves from being shot at by us. Mongrels! We now had no choice but to head south as quickly as possible. We couldn't risk wounding or killing our own men.

We soon found ourselves facing another predicament. The only bridge across the Aliakmon River had been blown up by another unit to slow the German advance. From our position I could see no way of getting across the river. Surveying the situation and finding no shallow ford where I could organise the men to form a human chain to cross, as RSM, I volunteered to swim the river and get to Brigade HQ and alert them to our situation.

After safely swimming across the river, I contacted HQ. At HQ the engineers managed the river crossings. Our job was to hold the Germans off for a few days so this would be built.

I swam back across the river to join the men and assessed the situation. We still had anti-tank rifles and by being conservative with our ammunition—only shooting if we could get a clear shot—we stood a good chance of holding them off long enough for the bridge to be built. We were successful in knocking off a few tanks early, so thankfully they left us alone after that. I managed to get what was left of the battalion across the river, via the unfinished footbridge our engineers had made, which was destroyed as soon as the last man crossed.

The men and I moved on to Bralos. I noted in my diary how we had lost quite a few good men and after my good friend Charlie (Charles Henry Jewell NX5996) from Wagga Wagga was killed, I felt I had had enough. I remember thinking, 'Let them have it. It's us or them.' We made it back to Rail Head and secured ammo for ourselves and the 2/1 machine gun lads with Douglas Channell. These men arrived almost at the same time as us; it was a good feeling to have more of our own with us. Most of the men had left by the time the last few of us departed with the Germans just over thirty yards away.

The thing that stands out most in my mind of this time is an elderly Greek peasant woman sitting by the side of the road. Her clothes were ragged and she was very thin; but in her hand she held a small piece of bread. As we passed she smiled and offered us her only piece of food. She'd lost everything and was starving, but she would have given that last piece of bread to an Australian soldier. It sent a powerful message to all of us there.

At 0830 hours the next day we marched further east to take up our next position ten miles east of the harbour. After camp setup was completed, including hiding our gear in the surrounding olive groves, I headed to the beach for a good cool dip in the clear blue sea. It was a great feeling to bathe and wash off days of grime but more importantly taking the time to clear my head. There had been too many times lately where this small luxury had been denied us.

Next morning reveille sounded at 0700 hours. We endured some air raids and bombings over the harbour. With only two hours' notice to move, we were off to board the destroyers and embark to Heraklion, the main town of Crete.

Costa Rica ship

We managed to board the *Costa Rica*. The naval evacuation to Heraklion was long and very hazardous for the convoy of ships. As luck would have it the Costa Rica was bombed. Fortunately I managed to scramble aboard a destroyer, which drew up beside the damaged ship rescuing many of us before the Costa Rica sank. Some poor devils missed the deck and fell between the ships. They were either crushed or drowned. I cannot speak highly enough of the sailors of our wonderful Mediterranean Fleet. They were absolutely heroic in their actions to get us to safety.

Greek Government concedes

On 21 April, with the situation quickly deteriorating the Greek Government realised further resistance was futile. Their army, which had been able to drive back the Italians, had now been cut off. As the government was unable to obtain further help by way of soldiers or supplies for these men, their only option was to capitulate.

The Greeks requested that the British forces be withdrawn. It was now merely a question of continuing a dogged withdrawal. The Greek Government's attitude was one of simple logic. *"You have done your best to save us. We are finished, but the war is not lost. Therefore save what you can to help win the war elsewhere."* It made us realise just how truly honourable these people were. We may have arrived as strangers but these people embraced us and took us into their hearts, we would leave as lifelong allies.

Just like the Greeks, every Allied unit had given its all, standing side by side and fighting against a common enemy. Many of our friends remained buried in the cold mountains and along our escape routes, many more were captured and taken as prisoners of war. Thankfully, tens of thousands of Allied troops made their way by ship to either Crete or Alexandria. What would the future hold for us now?

Maybe!

(Alan Eason)

I knew, with the evening shadows falling around them and regardless of the fact one of the rescue vessels was stranded, there was still a chance to escape Greece by boat. It was the evening of 24 April 1941, around 150 nurses from Australia and New Zealand, non-combatant soldiers from Headquarters of 1st Australian Army Corps, various units from the base, ANZACs from 2nd and 6th Divvies, cartographers and some of the Heavy Anti-aircraft Battalion all waited to board one of the five ships that had made it safely to port. I was so relieved to get aboard safely. Once we had as many as possible aboard, the ships sailed off in a convoy headed for Crete.

Maybe not!

(Dr John Borrie)

Some medical staff and I were tending to the wounded in a makeshift hospital hidden in the olive trees at Megara. Many soldiers streamed by, leaving the sick and wounded men who could no longer walk with us. We did what we could to patch up the wounds and get these men to the safety of the olive trees where they could at least get some well-earned rest.

On the evening of ANZAC Day 1941, an orderly came running into my tent calling out, "Dr Borrie, Dr Borrie we need to move the wounded closer to the beach." Our only chance of escaping by sea had arrived. All the units formed into groups of fifty and at fifteen-minute intervals they moved towards the beach half a mile away. Weary though we were, we were heartened at the thought of boarding a ship back to Egypt and so we sang as we walked. Before it was our turn to board the ship, full to capacity, sailed off. No one left behind spoke, what could we say.

At 3 am a British major instructed us to make our way to Corinth, twenty miles on, and reach it before dawn. This gave us two hours. We were instructed to cross the canal and follow the road on the left up the hill.

Luckily our ambulances had not been sabotaged, so we worked together to load the seriously ill patients and sent them onto Corinth. I was surprised and grateful to see an older cousin of mine, Andrew Borrie, who was an ambulance driver with the 5th Field Ambulance.

Andrew had put his age down by five years to thirty-five when he enlisted. He had been too young to enlist in the First World War with his brothers and had been determined to enlist in the Second World War. We hailed from a large family, who had migrated from Scotland to New Zealand in 1852. Many of our relatives had served and Andrew, like my brother Alexander (17781) and I, wanted to fight for his country. The last I saw of him was when he drove off in an ambulance full of wounded. I wondered if our paths would cross again.

It was now up to the rest of us to get moving. I hefted my primary medical book, my surgical haversack and my typewriter and instructed the men to commence walking to our planned evacuation point. There was no singing now; we were tired, hungry and sick at heart. We had come so close; this would be our last chance. By the time we neared Corinth, the Germans were already attacking the city and blocking our pathway to freedom. Any hope of retreat came to a swift end as the ever trigger-happy German gunners were strafing our rear. We did the only thing we could; we hid and prayed. Before long we were surrounded and taken prisoner. It was Saturday 26 April 1941 and all semblance of freedom disappeared.

CHAPTER 5

Calm before the storm

Reprieve

(Ronald Collins)

The fortnight spent on Crete was to be the first and only time in my military career when I was not required to parade or do daily drills. The unit helped with the scavenging of guns from the ships that had been sunk or lay damaged in the harbour. Luckily there were some anti-tank guns on the island and these additional guns retrieved from the sea were a godsend. It was not easy moving them, especially as some of this work had to be undertaken in and under the water. Once stripped down it was easier to transport them to shore in the small boats the Cretans had allocated for the job. The ammo was easier to move but in some cases more difficult to access.

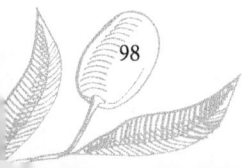

Members of 2/1st Tank Attack Regt.
on arrival back from Crete – May 1941
Standing - Roy Shepherd, Darcy Evans, Ron Collins (Splinter)
Ken Foote, Glen Scott, Eddie Boyd, Tom Roberts
squatting – Johnnie Merrill, Bill Madden (Sel)
Laurie Rulock, & Small.

> *Group arriving at Alexandria from Crete*

Although the Germans had spy planes in the air, there were only three or four raids during this whole time. Despite the fact, they inflicted damage to the bay area, our gunners were able to bring down three or four planes each time. I think this slowed their attacks and for the most part, my time in Crete was peaceful. The evenings would find us enjoying a meal and a drink or two in one of the local cafes. The Cretan people were wonderful and, just as in Greece, we soon formed a strong respect and friendship with the local communities.

At the end of the fortnight, the powers that be in Egypt reckoned as 2/1st Anti-Tank specialists, those of us from the unit left on Crete would be better off back in Africa. So when an ancient yacht, the *Lemnos*, arrived in Suda Bay with a load of coal it was requisitioned to take us to Egypt. The coal was hastily unloaded and without taking time to hose or sweep the decks, Stewy, myself and the others were shoved aboard amongst about two inches of coal dust. Stewy and I felt like sardines we were packed together so tightly. Boy were we pleased when we arrived back in Alexandria. We were covered in coal dust from head to toe and the first thing we did was hit the showers.

At the time we did not realise the full extent of our luck, embarkation from Crete happened on 19 May 1941; Operation Mercury began the following day.

Stewy told me that Mrs Stuart's little boy's guardian angel was certainly looking over him that day. In a letter Stewy sent to Weipa on 25 May 1941, he wrote:

> *Darling,*
>
> *Well I suppose you are wondering what happened to me as I haven't written for over six weeks but no doubt you have received my cable by this or at least I hope so.*
>
> *Well I suppose you have read all about our affair in Greece so there is very little I can say about it. I was one of the lucky ones and got our [sic] very lightly although I thought Mrs Stuart's little boy's number was called out several times.*
>
> *Greece itself is a very nice country but at present or while we were there it was a little too hot for our health conditions. We had a very frigid Easter. Guess what happened besides seeing the (…) above all things it snowed the first time and I also hope the last time I've ever seen snow.*

COMMONWEALTH OF AUSTRALIA—POSTMASTER-GENERAL'S DEPARTMENT.

RECEIVED TELEGRAM

URGENT RATE.

Telegram sent after evacuation from Greece

It fell for about three hours without a stop and then again later on in the night. If it had been under different conditions it would have been really enjoyable. It looked nice falling but boy was it cold. It went straight through me. You see we were very high up somewhere near the Yugoslav border and Greece is all ranges and passes. The passes were our downfall as far as bombing was concerned. They used to wait for the convoys every time but on the whole he did very little damage with his bombs. Have you ever heard a siren on those cars? Well that is what a screaming bomb sounds like and you think it's a dead cert to fall just where your hole is.

The Greeks are a very poor race of people. Their biggest treasure when we were going up was petrol tins. There was no need to hide them from the air you hardly emptied them and they were down on them but when we were coming back they would hardly look to them. They were down on all the preserved meat, jam and margarine and tinned fruit which needless to say we had our share of. We came across a big RASC food dump in one place the other side of a place named Larissa and there were tinned pears, peaches, apricots, cherries, blackberries etc. and we lived on these with cream every meal and there were plenty of meals as you guess. It was good while it lasted.

Well our big retreat as we were on the move nearly all the time then we got off on naval ships which was the best thing that happened to me for weeks. Well we did a twelve hour trip to Crete. Thank heavens we have a navy. They treated us very well when we got on at about half past two in the morning.

Well we were on Crete for a fortnight which was a good spell. Very quiet I think we had about three raids the whole time and every time there were no fewer than three or four planes brought to earth by our guns. It was great to watch them coming down and everybody cheered when one started to smoke.

It was on this island that I met your brother. He was camped a few miles from where we were. I was lost at the time as our regiment had moved to another part of the island, not that I minded such a lot, anyway I was talking to some chaps out of your brother's battalion so I went back with them and it was close to midnight before I found him so I stopped the night.

You see we had got that way lately that we can sleep anywhere. Anyway he told me about his party with Jim in Benghazi it must have been a beaut. He said something about Jim being missing. Is it right? Well I didn't see him after that. I think he is still there. I might catch up with him somewhere else one day as I think there is going to be plenty of time.

Well from Crete we journeyed to Egypt in a ship which had been used as a coal barge. Do you think we were in a mess when we go off it? There were about eight hundred of us on it and it was about a 1400 ton Greek ship and everyone from the captain to the deck hands had a go at running it. Every time one of the escorting ships would run a message up they used to get a book about a foot thick and read the message that way and when they pulled the cord to blow a whistle coming into port it wouldn't go. I think the sparrows or a seagull must have built their nest in it. I got a few snaps of when we got off the ship before we had a wash and a shave. If they turn out any good, I will send some to you and let you see what a "Grecian Harrier" was.

. . . .

Lots of love

Stewy

No supplies

(Alan Eason)

Being attached to 1 Corps Troop Supply, meant I was part of the advance party and lucky to be on one of the first ships out of Greece. Normally when we arrived at our destination it would be our duty to make arrangements for those who were to follow. Depending on which section you were in you would either organise transport and fuel, or tents, linens, food and the delivery of mail to the troops.

Even though I was evacuated in the first ships headed for Crete, my unit were forced to leave a lot of our gear back in Greece. I was assigned to travel with the limited stores we were able to get on board. Inevitably, the stores always appeared to travel and arrive at their destination intact if accompanied by an attendant. I didn't mind, it was an easy job and the men on the ship were so worn out they just wanted to grab a spot on the deck and get some sleep.

Unfortunately this meant we arrived in Crete with little equipment or supplies. Add to this, the thousands of soldiers due every day over the coming week and we had a real problem on our hands. There was no way we could meet the demands with our limited stocks, even with the small stock supply already on Crete. There was a slim chance of more supplies arriving from Greece or perhaps an influx of supplies arriving from Egypt.

As the days followed one after the other and the supply ships from Egypt were bombed and destroyed it became evident to us we had to do with what supplies we could source from the island.

Emotions ran high, but I especially felt sorry for our combat troops as they arrived worn out, shocked, hungry and in many cases in uniforms they had been wearing for days or sometimes weeks. Many of them had only their small arms, some ammunition, a ration pack or two, their great coats and a blanket. It was hell not being able to have tents, beds, clothing and food ready for them when they landed.

Many of the men were suffering from physical wounds, others suffering mentally from the effects of the war, depression due to our forced withdrawal, and the uncertainty of knowing what had happened to mates. Combine these factors with the loss of our vehicles and supplies and you can only just start to imagine the tremendous hopelessness we all felt at this time. We struggled between our feelings of pride and patriotism for our cause and the conflicting feelings of shame, guilt and betrayal at having to withdraw from Greece.

When you join the services, be it army, navy or air force, you instantly become part of a combined group. All the men in your unit become your brothers in arms. You live, work and breathe experiences that are unique to your group. You depend on the man next to you having your back. You work as a united entity to achieve your assigned objective.

Some of the guys and I took some time to wander the streets of Crete trying to source food and shelter. In the evenings we would connect with the locals, grab some tucker and drink some wine. Our unit was able to get billeting for quite a few of the officers and men but there were just too many soldiers and not enough housing.

Some of the locals were instrumental in helping our hard working medical teams get our sick and wounded housed first.

We heard a couple of our nurses had been injured during the evacuation but after having their wounds attended to, they kept working. We were constantly amazed by their resilience and determination to tend our sick and wounded especially after what they had already been through.

On the docks one morning I met up with some members of the New Zealand Medical Corp who had made it to Crete. One of them was a guy called Andrew Borrie. I helped him organise the sick and wounded he had transported on the final stage of the evacuation out of Greece into the hospital on Crete.

The people of Crete welcomed us with the same gusto as the Greeks. Initially it was difficult for these people who lived a simple life, restricted to their villages to comprehend why we would travel half way around the world, leaving our own families, to come and help them save their homeland from invasion. This had not been the case in years past when others had come to invade them, not help improve their lives. The bonds of friendship we built with the Cretans during these vexing times would last a lifetime for most.

Very quickly we came to learn the Cretan people are a very proud race who live by a very strict code of honour. Despite thousands of years of occupation by various cultures, the people of Crete did not cower or assimilate to the various occupiers' cultures. They remained true to themselves and maintained their own culture and religion. Cretan society is a closed society, firmly based around family, property and those living within their village. Their ideology is based on hard work and belief the land will provide for them. While wary of strangers, once you have been taken under their wing, their strong social code makes you a friend for life.

This ideology made working with the Cretans that much easier. You knew exactly where you stood with them. The strong sense of pride, belonging and honour the Cretans held was something the Germans failed to recognise. They were to pay a heavy price for this misunderstanding, especially in the weeks to come. The Cretans courage and loyalty was something we greatly admired.

Early British preparations

At the start of the Greek campaign, the 5th Cretan Division had been directed to travel to northern Greece to help in the fight against the invaders. This left the British to defend the island of Crete. The 14th Brigade initially sent to Crete to improve the defences from 11 November 1940, had achieved very little in the preceding six months. For reasons known only to those in charge in the War Office in England, the British had appointed no fewer than six different commanders in this short period of time. This may have come about because of the bombing London had been experiencing over the past twelve months; who knows!

It is no wonder the men under their command accomplished very little in the way of defences on the island. They did manage to start construction of an airfield in the west of the island at Maleme but this was of little use as there was no active air base operating at any of the three airfields on Crete.

Added to these highly inefficient results, the expected reinforcements from Egypt were not forthcoming. They were engaged in fierce fighting in the Middle East and this also ensured no aircraft were free to be based on the island. This left Crete undefended by Allied aircraft.

Misleading communique

In a communication dated 1 May 1941, Major General B. Freyburg stated:

> *The withdrawal from Greece has now been completed. It has been a difficult operation. A smaller force held a much larger one at bay for over a month and then withdrew from an open beach. This rear-guard battle and the withdrawal has been a great feat of arms. The fighting qualities and steadiness of the troops was beyond praise.*
>
> *Today, the British Forces in Crete stand and face another threat, the possibility of invasion. The threat of a landing is not a new one. In England, we have faced it for nearly a year. If it comes here it will be delivered with all the accustomed air activity. We have in the last month learned a certain amount about the enemy air methods. If he attacks us here in Crete, the enemy will be meeting our troops upon even terms and those of us who met his infantry in the last month ask for no better chance. We are to stand now and fight him back. Keep yourselves fit and be ready for immediate action. I am confident that the force at our disposal will be adequate to defeat any attack that may be delivered upon this island.*
>
> *Source: "Special order of the day,"* by Major General B. Freyburg, V.C., C.M.G., D.S.O. Commander, British Troops in Crete. Australian War Memorial

At the start it may have appeared the one thing we did have in our favour was the strength of personnel stationed on the island. Those originally posted to Crete, combined with the arrival of troops evacuated from Greece increased the numbers on the island two fold. However upon re-evaluation it soon became obvious that while we had the numbers, over half of the soldiers were not combat troops. They were supply troops, assorted personnel and a large number of prisoners of war. Those who were combat troops felt pretty useless without the bulk of their heavy guns, arms and ammunition. The harsh reality was the fighting soldiers on the island were ill equipped for the forthcoming encounter with the Germans. The Cretans had no military equipment and it was unclear early in the piece if any soldiers from the 5th Cretan Division, who knew the island so well, had escaped from northern Greece.

Formed into working groups, many of the non-combatant troops were assigned to help with the salvaging operations being conducted at Suda Bay. A number of warships, troop and supply ships had been sunk or left burning in the harbour after some early air attacks by the Germans. The Cretans provided small boats enabling us to salvage what we could from the numerous disabled ships. Men would go out to the ships and strip them of any useful guns, ammunitions, food, blankets, fuel and any other items deemed useful to defend Crete and help with the welfare of the troops. Of the 2000 tonne of supplies shipped from Egypt only around 200 tonne made it safely into port.

The Australian Hellenic Greek

(Jim Zampelis)

The chief cook came up to me and said, "Zampelis, you speak fluent Greek, go and source a building where we can operate the kitchen or organise sand bags between two buildings and create a space so we have somewhere to work from." I immediately took off to carry out his request. After talking with some of the locals I was able to find a small empty shed and this is where we set up.

I prided myself in providing the officers with the best meals possible under such trying circumstances. Where possible I would source fresh food daily from the local markets and farms. Once again my Greek heritage and knowledge of the Greek language worked in my favour. However it didn't matter how you looked at it Crete was not set up to cater for the volume of soldiers now entrenched on the island.

It was clear to see life was extremely difficult for the Cretan people whose only means of support was earned from a meagre living off the land and trade with Greece. In normal times they would export their excess produce, which included olives, olive oil, grapes, tomatoes, cucumbers, zucchini, potatoes, oranges and carob beans. The carob seed was ground down to make flour, which was used in many of their foods. In addition to this they also produced a variety of herbs, honey, some nuts, yogurt and edible meats from their sheep and goats.

In return they would import from mainland Greece the basic items they were unable to produce on the island. As the war progressed, importing and exporting was impossible with all the bombing of ships going on.

This played to our advantage and the Supply Corps purchased these goods from the Cretans to feed our soldiers. They were grateful for the income and we were desperate for the supplies of fresh food, no matter how small the quantities. Every morsel counted if we were to maintain our health for the coming encounter.

Building materials like stone, marble, lime and building blocks were also processed on the island. From what I saw their craft industry consisted of a few ceramics, soap making and leather-work.

Unfortunately for us, all of their fuel came from the mainland and ours had to be shipped to the island, so once the current supplies dried up there would be no more, the few vehicles we had were rendered useless.

The sick and wounded

(Andrew Borrie)

As soon as I disembarked in Crete I was assigned to the 7th British Hospital. Being an ambulance driver meant I had plenty to keep me busy. I transported wounded and sick troops and sometimes even injured members of the local population to our small hospital. It was here I came to witness the amazing work of our medics. The doctors worked inhumane shifts at times and the masseuses were never short of work.

It defies belief just how courageous our nurses were. They were everywhere. Working beside the doctors, listening to the seriously wounded, calming the dying, they became mother, sister, sweetheart and scribe to all those who needed them. Their calming influence, care and concern for the men was immeasurable. I swear sometimes it was as if I was witnessing a miracle at work. Never could I speak highly enough of the work they carried out.

I assisted the medics dealing with some horrific wounds sustained by the men. Some of the worst were shrapnel wounds, gangrene, burns and gunshot wounds. Down at the dock I was required to help rescue and treat sailors and soldiers from the ships that had been sunk in the harbour. Many of these men suffered horrific burns and shock was a major factor in whether they lived of died. The sooner we could transport them to the hospital the better.

One man I found was hidden beneath a small upturned boat. It was sheer chance that lead me to lean on the side of this boat on the beach to take a spell, but it was the smell that first attracted my attention. The sailor had been wounded and suffered terrible burns to his body when the ship he was on had been sunk in the harbour. Having made his way to shore towards the evening, he had managed to crawl under the boat trying to gain some protection. Unfortunately the next morning he was too weak to crawl out and call for help and lay semi-conscious and undiscovered until I found him quite by accident. By now it had been two full days since he was injured and his wounds were covered in maggots. I called out to one of the other chaps and between us we placed him in the ambulance and got him to the hospital as quickly as possible. Though every effort was made to save him, he passed away later that evening.

Heading for Retimo

(Douglas Channell)

Basically the plan was a simple one, defend the three airfields on the island. These would be defended by New Zealand at Maleme, Australia at Rethymnon while the British would hold and defend Heraklion. However not all the soldiers of these nationalities were together, they were divided between various groups and placed at other spots around the island. Australia for example had the majority of their men at Retimo, with others at Suda Bay and Georgioupolis, while headquarters were camped at Khania. Not the most ideal situation.

On the home front the Cretan police and cadets took up arms, with the 1st Greek Regiment (Provisional) combining with many of the armed civilians to help where they could.

My unit and the 2/1st Australian Infantry Battalion were dispatched to join the garrison at Rethymnon. In true Aussie style we shortened the name to Retimo. This garrison consisted of men from the Greek 4th & 5th Regiments, 2/11th Battalion, 2/8th Field Company, 2/1st Machine Gun Battalion and the 2/7th Field Ambulance. Sixty miles to the west of us, two New Zealand brigades were holding the main aerodrome at Maleme. Forty miles to the east, a mix of British and Australian forces were entrusted with a similar task at Heraklion. We were all tasked with the same objective. Don't let the Germans gain control of the airfields.

On receiving orders to advance my unit to Retimo, we set off under the command of Lieutenant-Colonel Ian Campbell. Once again the unit travelled on poorly made roads, which were little more than tracks up and down hills and valleys. It was almost impossible for trucks to traverse and very hard going for the men. Thankfully, it was only a short journey this time. The Cretan landscape was very harsh and unyielding making the journey seem even harder and longer. To make it easier I started singing and encouraged the men to join in as we marched forward in formation. It may have been hard going but we were making light of it in the best way we could. I have always been a firm believer in humour and encouraged my men to indulge whenever the opportunity presented itself. Some outrageous ditties caused much laughter and everyone felt all the better for the typical Aussie use of humour in dire straits.

The city of Retimo is a very old town with a 19th century lighthouse proudly standing in the harbour, an old fortified Turkish fortress, a number of taverns, a harbour filled with fishing boats and a long beach. The architecture was heavily influenced by a Venetian style left over from an earlier Italian occupation. There was an olive oil factory just out of the city where olives from the surrounding farms were processed. None of these factors made the city itself an object of military importance, however, it was the emergency landing strip seven miles to the east of the town that was deemed of vital importance. For the foreseeable future it was up to us to defend the emergency landing strip at Retimo. This important airfield must be denied the Germans at all costs.

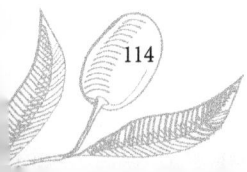

My men formed A Company and we waited in the early morning just outside Retimo. Camp was set up below the low hills overlooking the airstrip and work proceeded to outline the best positions for our defences. That night everyone slept out in the open with only a ground sheet. If you were fortunate enough to still have a great coat it went a long way to keeping out the cold night air. The following day dawned clear with cloudless blue skies. After reveille, the first job of the day was to start digging slit trenches and weapons pits. Digging was tough because the ground was so stony and we had very few picks and shovels at our disposal.

After Major General Freyberg VC came and inspected our progress we were soon issued with blankets, clothing and digging implements. Progress on the weapons pits improved measurably after this and the men detailed to carry out the camouflage did so with great gusto and ingenuity. Their training really came to the fore here. The pits were ready but empty as I had been ordered to leave them unoccupied for the time being.

Lt Col Campbell was very pleased with our progress and continued to organise the sighting of each of the weapons pits making sure the entire length of the airstrip was covered from Hill A to Hill C. Next came the task of digging tunnels from the back of the hill to the pits. As we were living and sleeping in the slit trenches at the rear of the hill, we required access to the pits from the rear to eliminate the possibility of exposing our positions by entering from the front.

The runner

Very early on in the piece I came across a young lad called Marcos Polioudakis, a native of Retimo. One evening as I was heading back to the dugout, I was keeping watch on what was happening ahead of me. Each time we ventured out I would have the men return by a different route. I was training them to be wary of being spotted by any Germans who might descend on the island. On my way back this particular night I came across Marcos. He was walking a short distance in front of me. I hailed him and we stopped to talk.

He was a young man, thirteen or fourteen years of age, on the cusp of manhood. By his reckoning he was ready to fight the Germans to keep his village safe. As we walked back to base he reached into his shirt pocket took out some papers and handed them to me. He had spent time drawing mud maps of the local area, marking tracks, caves and good vantage points. He also told me the names of the important people in the village. I quickly came to the conclusion this young man would be very valuable to us and made him an offer to work with us. His eyes lit up and he was overjoyed to know he would be a very useful contact for us. He quickly accepted and became our runner and water boy.

Marco was to prove his worth many times over. His bravery was astounding for one so young. He and I formed a strong bond during the short time the 2/1st Infantry Battalion was stationed at Retimo. Working primarily as a water boy, he also delivered messages and found food sources for us. Though we had limited combat rations available to us, the locals were happy to sell us fresh produce, which enabled us to enhance our food intake. There were many other small jobs Marco undertook making him an invaluable asset to our small team in the coming days. Marcos became the major link between the unit and the Cretan people during the Battle of Retimo.

A small portion of the 2/3 Field Regiment now equipped with old French rifles and very little ammo, together with two untrained and poorly armed Greek Battalions joined us. The coming encounter was not going to be a walk in the park, though we did have very specific instructions on how to fool the Germans into thinking the airfield was only defended by a few ill-trained Greek soldiers. To this end we dug dummy pits by the sides of the airfield and stayed out of the camouflaged pits, all the while hiding from the ever-watchful Hun planes. When the German reconnaissance aircraft flew over taking pictures of the ground defences they had a totally misleading take on the situation.

Gathering everyone together each day after reveille, I instructed them in drills, defence and surveillance of the surrounding areas. Marco's maps were very helpful in this regard. We dug ourselves in deep at Hill A, making ourselves invisible from prying eyes above. I had selected the best positions available for our defences and hoped to heaven we could pull off a miracle.

During leave the men were allowed to go into town five at a time. Pickets were posted in Retimo and the closer surrounding villages. Even though there were no tents or other such comforts, one of the most enjoyable aspects of being at Retimo was the ability to go swimming. It was wonderful to jump into the sea after a hard day digging. After a swim and a good wash down with soap one felt like a new man. The sea was also a place for us to wash our clothes. A great time to get a little bit of sunbaking in while the clothes dried. Simple but effective!

Marcos worked very hard sourcing food and information from the people in his village. These people though poor in wealth were intelligent, literate people. They knew the dangers they faced and while they had very little themselves, they were generous to us.

In the days leading up to the invasion of the island, village life continued on. The villagers could be heard in the evenings playing the lyre and the lute, singing and dancing. The older women of the village would sit in the early evening weaving or doing needle crafts while the older men worked on their leatherwork, wood carving or simply enjoying a drink and a game of cards.

Soldiers and civilians had prepared to the best of our abilities for the forthcoming battle. Most of the people we had encountered were good respectable people, but there were bands of 'bandits'. This was the case the world over and it was no different here. Some of these men later joined with the villagers and became members of the resistance.

Dreaming of Menzies

(Alfred Carpenter)

The 2/4th Battalion landed in Crete at 0400 hours and proceeded to march to the outskirts of town just off from the main airfield and set up camp. "Carpenter, the CO wants a word with you, now." I headed to the CO's tent, "Alf we need to spend some time working out how we should proceed from here."

We had lost quite a few men during the Greek offensive and some were in the hospital receiving treatment. Our transport section had no vehicles or heavy artillery. So the day was spent working out how to regroup, arm and have our men as battle ready as possible for the assaults we knew were to come.

At the end of what turned out to be a long day, it was good to have three of the sergeants from the 2/3rd join us for a drink. I must say we did make a bit of a night of it.

Stand to at 0530 hours was a little tough after such a big night but it was payday so no one minded. Our force had increased as one lot of the Australian Service Corps Units, 2/2nd Battalion and 2/3rd Battalion were attached to us, together with fifty New Zealand chaps.

A couple of my men who had missed the ship we evacuated on eventually re-joined us in Crete. They relayed their story of escape. After our ship had slipped off from Greece into the coming dawn they had found a fisherman in one of the bars who hailed from Crete. He agreed to take them to Crete for a couple of bottles of whiskey. It was an adventure far removed from their trip from Egypt to Greece that's for sure but we were very pleased they arrived in Crete safe and sound.

My battalion located to Heraklion above the airstrip around and on the two hills known to us as 'the Charlies'. We had troops on the hill furthest away from the airstrip and, during my patrols of the area I noticed some very old ruins. They appeared to have murals painted on them. The bombing by the Germans destroyed some of the site and covered what we determined were old graves.

Over the next few days I got some time to myself and managed to write some letters home. Around five percent of the men were allowed to go on leave at any one time to spend the day in Heraklion. This allowed them to have a look around and make some purchases. These consisted of gifts for home or food to supplement their rations. Those not on duty took the opportunity to swim and sunbake while the Germans were being unusually quiet.

Clear weather conditions and glorious sunshine allowed us to start building a bomb shelter or funk hole as we called it. By the afternoon I was making provisions to establish Battle Headquarters. While digging the funk hole in the shale in the side of a cliff we saw an enemy spotter plane. We quickly downed tools and hid until it had passed. As we were working without our shirts on we all soon ended up with good tans. No comparison to the icy cold weather we had experienced in the Greek mountains only a few weeks before. I was so pleased that was well and truly behind me.

The following morning dawned overcast and looked like rain. We spent the day digging the funk hole. Some of the guys had got together and produced a play, which we gladly attended. It was an Air Force concert and we all donated to the Spitfire fund. The next day was payday so I dropped what money I had left into the collection hat.

Another quite clean day today, no German presence in the sky. After reveille I joined the men who were digging out a tunnel. This task needed to be finished by the following night when we were to swap positions with Black Watch. The task was finished on time and the swap over commenced at 1900 hours and was all completed by 2200 hours.

The Padre had located a small cave nearby where he set up for mass. Church parade was called at 0900 hours and many of the men took the opportunity to attend. There were no religious barriers here, we all prayed together.

A bonus arrived later that morning in the form of a clothing issue. My uniform was definitely the worse for wear and it was good to have a swim, a wash and be able to kit up in some fresh clothes. Whacko! I sure felt a million dollars after getting cleaned up. Time to go have a drink.

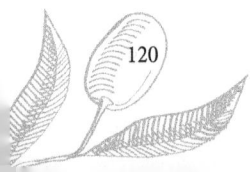

In February of that year, while I was stationed in Benghazi, General Blamey accompanied by the Australian Prime Minister, Mr Menzies, met with some of the units. We were at attention but must have looked a rag tag lot. Battle-scarred, stained and in ragged uniforms, many of the men had long ago worn out their army issue boots and some wore Italian leather boots. Within days of their visit we were fitted out with new uniforms. It was little things like a fresh uniform or a new pair of boots that we take for granted in normal times that ended up meaning so much to us in times of war. I knew I would never take clean clothes or well-fitting boots for granted again.

The balance of our battalion stationed at Suda Bay arrived. The 2/4th landed in Greece with an 1100 strong battalion, sadly we were now down to around half strength, having suffered many fatalities and losing others who were taken as POWs.

Four of the men had decided to stay in town and indulge more than they should have and ended up being reported AWOL. This required me to go into town to the provost office and interview witnesses. The four men each received 28 days detention.

Early the next morning the Germans pounded us with continuous air raids. There was no rest for us as we had to be on our toes and on the move as orders came through to reposition a mile further on to join up with the CO.

The next day we witnessed a dog-fight very close to our position. One Gladys against five Dorniers. It was a spectacular show and our Gladys crew gave the Jerry's something to think about. Dorniers came over at dusk and dropped a load of bombs. Our guns scored a hit and one came down in a ball of flames. Didn't get much sleep as raids continued on and off all night.

Over the next five days I recorded the following notes in my diary:

15th May: *"On and off raids all night. Beautiful air raid. Heavy raid at dawn. 3 Dornier's shot down while writing to Marjorie had to duck for cover under an olive tree. Planes over during day. Evening heavy raid as usual".*

16th May: *"Stand to at 0500hrs with a beaut bombing raid on Greco Barracks. 2 Hurricanes arrived this morning. Whacko for Jerry now! Easy day".*

17th May: *"10 Bristol Beauforts arrived. Looks like quite a show. Jerry must get the word through about these crates, as there was no more of his planes over".*

18th May: *"Our planes left during early morning. Jerry over again with MG and bombing. Erected a trip wire across our wadi, let 'em come up it again strafing us".*

19th May: *"A couple of machine gunning raids. Am getting used to them, if it were at all possible. None of our Air Force left".*

By this stage, we had not washed for a week or two. Max, our CO, said to me: "We'll go down for a swim in the morning."

Arriving at 42nd Street

(Reg Saunders)

I was pleased to be able to stand on solid ground once again. It was invigorating for me to stand with plenty of space around me, free to move my arms about, stretch and breathe in fresh air after the nightmare trip on the ship from Greece to Crete.

Major Walker advised the 2/7th unit we had been ordered to head for 42nd Street, so named by the 42nd Field Company, Royal Engineers who had been deployed to Crete in November 1940. Their job had been to improve the road and in order to accommodate heavy army vehicles they used the 'sunken road' technique. The road was well built and the surplus dirt had been piled on both sides of the road. It was here the 2/7th joined up with a New Zealand unit already in position. We needed to dig in and prepare to defend this highly important road.

CHAPTER 6

Giant birds and silk clouds

A young boy's memory

(Jim Samios)

My family and I lived on the island of Kithira situated to the north of Crete. Even though I was only seven, I vividly remember the events that unfolded on that fateful day of 20 May 1941.

As usual I awoke, got dressed quickly and went downstairs for breakfast as usual. My family were eating when we became aware of a humming sound growing louder and louder. Quickly we emerged from our home to find out what was happening. It was a beautiful spring morning with not a cloud in the sky. Looking up and off in the distance I could see what looked like many giant birds heading towards us. There were hundreds of them.

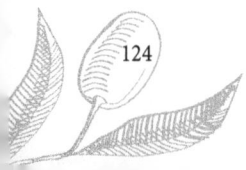

I couldn't understand what the big birds were doing so I climbed onto the roof of the small building beside our stables and watched them fly over all day. Every now and then there was a big bird surrounded by smaller birds. My daddy explained what I perceived to be big birds were in fact planes filled with paratroopers and the smaller ones were escort planes. These 'birds' were heading towards Crete.

Sometime before this, about a dozen British war ships filled with troops had docked on the island of Kithira. Our island was the first place the Allies landed on their way to Greece. My dad spoke very good English and they asked him to act as an interpreter. He was able to tell them for four years there had been no doctor or medical help available to the people on the island. Each day a car would be sent to our home and Dad would go to the port. Here the English opened the doors and offered medical aid to the islanders night and day for weeks. They were very good to our small island community.

Ships in port at Kithira

On the side of the mountain above our village a beautiful big church had been built many years before. Eighty percent of the soldiers wanted to go up to the church to have a look at it. The church was built in the rocks, stone and plaster. It had a big cross at the top and the whole building was painted white. You could see it from everywhere in the village. Being white it really attracted attention.

We had enough to eat, not too much, but enough. We grew all our own vegetables, as the soil around our home was very fertile. Everything you put in the ground grew. There was plenty of rain and sunshine, perfect conditions for growing our vegetables. We were lucky to live on Kithira where there was no military base and no Allied soldiers based there.

Living on this very small island was the only thing that saved us from the fate faced by so many Cretans. Our island had Italians on the north and Germans on the south. There was no conflict between the enemy and the villagers. We rarely saw the Germans as they had plenty of food but the Italians were not as well prepared.

As a little boy I loved the beautiful animals the Italians brought with them. They were mules—not horses—just mules. They would come and take some of the bread, wine and oil we had. They just took it but they left enough so we wouldn't starve. Every month or two you knew they would be back to take more.

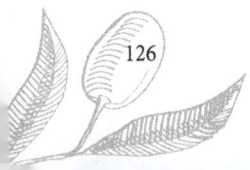

My Dad had to go and work for the enemy, sometimes two or three days a week. He would work, they would feed him and sometimes give him food to bring home for the family. They were fair and caused no problems with the villagers. They didn't kill anyone and they didn't interfere with the women.

I remember Dad was a very hard working person. He was very much against the Germans but he never showed his hatred to them. He did the work they requested and didn't argue with them when they took some of our food and there was no trouble. Every now and again though he disappeared at night. Where he went I did not know. Later I found out he and a group of friends went to watch the sea. If they found anyone in a boat who had escaped from Crete, they would hide them in a cave and bring them food. When they found a way they sent the majority of the escapees to Egypt. They had connections with other groups and the British and so they worked together to look after escaped Allied soldiers.

Every night, life went on as usual, except we could not have electricity and the curtains of the house had to be drawn all the time. Olive oil was the most important thing we had. We had a dish with oil and a bit of cloth, which we would light so we could have a dim light to see inside. We had to economise as the olive oil became scarce. Even though we couldn't have lights on in summer there was still dancing and singing in the street at night.

Kithira was never bombed but my Dad would gather us together and take us to the bomb shelter he had built in the back yard. During the war we lived downstairs as Dad was scared to live upstairs in case it got bombed at some time.

I remember many years later one of the soldiers Dad helped escape wrote thanking him for saving his life. By then he was safely home with his wife and children but he never forgot what Dad had done for him.

After the war the Germans and Italians cleaned up all of their armaments. There was no live ammunition, no landmines, nothing was left on the island anywhere that could endanger the lives of the islanders. Our circumstances were totally different from Crete. If it wasn't for the resistance on Crete the Allies may never have won the war.

The Enigma Code

In an operation codenamed 'Mercury', Hitler sent almost 10,000 paratroopers to the island of Crete in what he thought would be a surprise attack. However, unbeknown to Germany, the Allies had captured an enigma code breaking machine, which allowed Allied intelligence workers to decipher the German codes and plan military actions accordingly. While on one hand this information was used significantly to the Allies advantage, it also had to be used circumspectly so as not to alert the Axis powers. A fine line for the Allies to tread!

By breaking the German code the Allies knew the major objective was for Hitler's elite paratroopers to attack the defences at the airfields. This would then allow the German gliders to land and disembark land combat troops, weapons, food and other supplies. Hitler's troops had a three-day mandate to accomplish this outcome and to leave a small force of troops on the island. With that under his belt, Hitler could then recall the majority of his force from the island adding them back into his main troop, enabling him to concentrate once again on his planned invasion of Russia.

Even armed with this information, Lieutenant-General Bernard Freyberg, Commander of operations on the island of Crete, did not appear to believe the intelligence. He seemed convinced the Germans would engage in a seaborne attack and acted accordingly, spreading his troops in thinly scattered units. Primarily the New Zealanders would cover Maleme, the Australians were sent to Rethimno [or Retimo as we referred to it], while the British were sent to defend Heraklion.

Even though the Allies had double the number of troops, over half of them had arrived from the defeat and evacuation of Greece three weeks earlier. There were only around 7,000 fresh troops and a large majority were non-combatant troops. There was very little heavy artillery, machine guns, rifles, ammunition, food or motor transport available and most importantly no air support. Despite having three airfields on the island, there was no air force stationed at any of them. Freyberg had sent the few operable aircrafts on Crete to Egypt knowing they would be of no use in the upcoming confrontation. What the Allies on Crete did have in spades was courage and determination. They would do their utmost to deny these attackers their ultimate goal of gaining control of Crete.

Silk clouds

Douglas Channell

In the early morning, I waited just outside Retimo with my men of the 2/1st Infantry Battalion A Company. I had the men take up positions on the low hills overlooking the strip and we waited. It was not long before I received word the air assault to Maleme to the east of the island had begun.

The New Zealanders took a battering. Heavy strafing was followed by relentless bombing, the paratroopers started jumping out of planes over the airfield at Maleme. The New Zealanders held their positions and started to defend the airfield in earnest. Very soon I knew it would be our turn to face the Huns.

Michael Kennedy was with me on that fateful day of 20 May 1941. Michael wrote and I can verify his comment:

> *On the afternoon of May 20, the great German air offensive commenced. From the sea came an ominous roar. As the men in the olive groves strained their eyes they could discern the outlines of the approaching enemy host. First came the Messerschmitts, which tore over the treetops with their Spandaus blazing as they strafed the infantry positions on the edge of the aerodrome. These were followed by the inevitable Stukas, which moved high in the heavens with incredible slowness, to peel over on their backs and drop like plummets, with a whine which became louder and louder, until it was unbearable, ending with an ear-splitting roar as the bombs thudded into their targets. Finally, the troop carriers, which hovered like vultures, were overhead. Suddenly from the side of each carrier, a black dot appeared, to be followed by another, and yet another; then the air was filled with a swirling mass of colour, as red, white, green and brown parachutes cascaded against the blue Aegean heavens. At first they swayed crazily as they caught in the slipstreams, then steadying, fell earthwards.*
>
> "Retimo—A Lost Victory" *The Sydney Morning Herald*

Silver ghosts

It was not long before the German troop planes converged on our position trailing gliders behind them. Now it was our turn to witness what initially looked like hundreds upon hundreds of falling black dots. These dots covered the sky morphing into colourful swirling red, white, green and brown silk clouds. Over 1000 paratroopers were arriving to the east, west and south of our positions. We were being surrounded.

The silver gliders appeared like ghosts in the sky, silently ready to release ground troops by the hundreds. They were determined to wreak havoc on the island, its inhabitants and the Allied troops stationed there.

This was the first time in military history a situation like this had happened. We had no experience of this type of attack, no idea how to prepare or how to defend against this type of warfare. I could only hope the training and preparations the men had undertaken would give us the upper hand in the forthcoming battle.

CHAPTER 7

The Battle of Crete

Alf & the 2/4th Australian Infantry

(Alfred Carpenter)

The following notes are taken directly from my diary on events that happened at Heraklion from 20 to 26 May 1941.

> *20th May: Stand to as usual at 0500hrs.*
>
> *I got security all sorted ready for our much-needed wash, when the phone rang: "Colonel Campbell said he's in trouble at Retimo. German paratroopers have taken the Aerodrome and we might be the next lot." Within a couple of hours, the area of and around Suda Bay was under bombardment. German Stukas were pinpointing anti-Aircraft guns.*

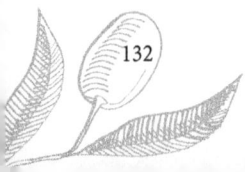

All of a sudden the attack stopped and we became aware of a throbbing noise. Taking up my binoculars I looked out to sea. I could see masses of planes approaching tier upon tier. They were headed straight for us. This sighting confirmed the earlier intelligence reports, Crete would be invaded by paratroopers.

1600hrs: IT'S ON! Whacko. Planes in 100s, smoke bombs and HE, MG. Let umbrella men come. Reg Johnstone killed. Paratroopers coming down in 1000s. Managed to get a Bren, fired it until barrel got too hot to hold. Gliders coming in, troop carriers shot down. Not many of umbrella men landed alive. There was a great tendency to put more lead into them than was necessary. Even Bob Talon, the cook, grabbed a rifle and said: 'I'm Tallon, the fighting cook'. All Battalion did their job nobly and well.

About 4 pm we heard the drones coming in at sea level, next thing when they reached land and gained height (for a drop from 120 metres) the paratroopers were landing all amongst us. I was wounded during the defence of Heraklion. Under a white flag, a German Medical Officer brought two wounded paratroopers to Btn HQ to be treated by an Australian Doctor. My fist instinct was to shoot the German MO; but under the Geneva Convention, this was not permitted. The CO concurred and the uninjured Officer was permitted to leave under the protection of a white flag. He returned safely to his men. Now, though, he could report on the Battalion's position, and sure enough, the following day, the company was mortar bombed.

The right side of my skull was hit by shrapnel and caused damage to the optical nerve. (I later had a corneal implant as resulting blindness affected my right eye). The Battle of Crete was hanging by a thread. Pockets of German paratroopers that had successfully landed on target, and airborne infantry that had been crash-landed into the hills, waited... The New Zealander's held their positions around Maleme, with their well hidden artillery now adjusted to shell the airfield. Another fierce battle started near Retimo with non-stop bombing from German planes. Six JU52s dropped supplies intended for their own troops. At Retimo, one of the defending patrols came across the body of a German Paratrooper with the colour signal codes used to message supply-carrying planes their requests. As plenty of German machine guns had been captured, the defenders of Crete requested ammunition to use; which was promptly dropped to them by (red) parachute!

Australian troops collected as much of the equipment as they could; Arms and ammunition (many of which were handed over to the fighting Greeks) desperately needed medical supplies, even motor bikes and a wireless set: tuned to German wavelengths. This proved to be a double-edged sword, as the reality of this situation was that the Germans dominated the sky. From the air, they were able to observe and maintain an efficient stream of supplies: an advantage that ultimately swung in their favour.

The evening of the first battle day on Crete sank with a red bloody glow.

22nd and 23rd May: *Got a couple of Luges, one prize Nazi flag and German compasses. Collected medical supplies dropped by Blenheim, taken into hospital. 16 Junkers over. Gear dropped into Jerry's hands this time. 11:20 Another bit of gear dropped west of town. Oh for a Hurricane or two! Jerry again collected 1700hrs. 40 Dorniers cleaned the town up. Bombs and MG. 6 Hurricanes arrived but left again.*

30 Dorniers dropped their bombs. The RAF acronym was changed to 'Rare As Fairies' by some Australian troops fighting in Greece. They felt there had been very little air support for them. The King of Greece and his government were taken off Crete by two British Destroyers. There was an acceptance that the island had been lost. The German Supreme Command dropped leaflets in their thousands, over Heraklion written in both Greek and English.

Naturally, the unexpected delivery of so much paper was put to good use by Australian troops on the ground. Having found the leaflets were not much good to use as cigarette papers, a more practical, toiletry, use seemed appropriate.

The expected news of success for the Germans stationed in Greece had not come in. It was only evident that up to that time 9000 men had jumped according to plan and 4 beach heads had been formed. But not one of the airfields, needed for the transport of airborne troops, was in possession. The General Command began to realise that the enemy's resistance on Crete was considerably stronger than anticipated.

24th May: *"0730 and it's on again. Bombs, MG and Paratroopers with stores. Whacko. Hurricanes down on aerodrome. Someone will get hurt here soon. Went through to aerodrome to get mortar ammo. Just 100 rounds. Sniper had a lash at us. MG and Bombers over all day.*

1700hrs and all well".

25th May: *"Continuous raid all day. MG and Bombers. Lester collected 100 rounds of mortar ammo. Through to ordinance to collect Jerry MGs, 4 of them".*

26th May: *"Awoke to the rattle of SAA. Plato's Platoon cut off".* *Australian and New Zealand battalions formed a defensive line along the Chania to Tsikalarion road south-east of Chania, forming a rear guard for the withdrawing troops. Although they were profoundly understrength, the defensive line they formed provided enough cover to not only assist troops in their withdrawal, but also to cause the Germans to withdraw for a brief period. ANZACs carried out a bayonet charge on advancing enemy troops. This action briefly halted the German troops.*

Our men were all lined up, ready to fight their way out. Sgt Swanson (Teddy) led the charge through. The noise of that charge was blood-curdling! I got a Brenn gun to give them covering fire and we got that platoon back to the lines again, so we reinforced each other. One lot of German paratroopers was in a wheat field, so we put petrol on the windward side and set fire to it. They came out with their hands up. If any of them fired, we fired.

At this stage the Germans on Ames Bridge were in a strong position and posed a very serious threat. Australian patrols continued to gather as much information as possible, but were told to withdraw if threatened, in order to best defend Heraklion aerodrome.

Next thing we got an order through dispatch rider to say we were abandoning the island. Our section was clear, the aerodrome was clear, but other strategic zones had been taken and now there were Messerschmitts landing additional troops. We were to prepare for a night withdrawal by ship at 3 am".

28th/29th May: *Under cover of darkness a force of 4,500 Allied troops were hastily but smoothly evacuated from Heraklion. Luckily for us, we were able to escape along the causeway and managed to board the waiting ships. To cover our escape and to fill our empty positions until we were clear, a section of string was rigged similar to what had been used at ANZAC Cove in the First World War. Rifles were set up with pieces of string attached to the triggers and cans of water filled to all different depths. Once the water level reached a certain level it would cause the string to pull on the trigger and the rifle would fire off a round giving the impression we were still occupying our positions. It would be hours before the Germans discovered we had flown the coop.*

A squadron of ships was waiting off shore. "We got aboard HMS Imperial & Hotspur, but then found one platoon hadn't shown up, so our Colonel managed to get the ship's captain to wait until 3:30 am. That half an hour made all the difference, as our stragglers arrived so then we left on the ship.

Now though, the sun was coming up; dive bombers started coming down on us. That extra half an hour of waiting for stragglers made our evacuation by sea even more fraught with danger. That's what the captain had been trying to avoid. The Luftwaffe remorselessly attacked the Fleet.

"Hotspur was packed full of withdrawing troops, there was standing room only. We still had our weapons though, so any bombers that came at us got everything we had.

The captain of Imperial commanded the crew to keep zig-zagging in the water to avoid the bombs being dropped, but one landed so close to Imperial, she was lifted out of the water. The boatswain reported damage and we were taking water fast: her steering gear broke down and it became clear that Imperial wouldn't make it to Alexandria. Semaphore flags relayed an urgent message to nearby Hotspur. She came alongside and we jumped ship. Some men missed the jump and drowned".'

Imperial was torpedoed by Hotspur to prevent her being used by the Germans. 'Shortly before noon, two Fulmars of the Fleet Air Arm made contact with the Squadron, and, although a few more attacks were made, they did no damage. Himself wounded, Admiral Rawlings of the Orion brought his shattered squadron safely back to Alexandria at 2000 that night - just 26 hours after we had sailed. Orion had two rounds of main armament left, no anti-aircraft ammunition, and only ten tons of fuel. She had lost her Captain, eighty-two of her ships company and some three to four hundred of the soldiers we had snatched from the hands of the enemy. Orion had filled the breach'.

Along with the survivors of the 2/4th Australian Infantry Battalion, I arrived at Alexandria on 30 May 1941. I felt a great sigh of relief. This had been by far the most horrendous experience in my military career to date. During the last month we had lost over 700 men, the evacuations from Greece and Crete had decimated our unit.

The Battle of 42nd Street

(Reginald Saunders)

Early on the morning of 27 May the Australians and New Zealanders formed the rear-guard for the retreating Commonwealth forces heading southward toward the evacuation beaches at Sfakia. The chosen position for the rear-guard action was 42nd Street. The street had been nicknamed by the 42nd Field Company Royal Engineers who had been deployed to Crete in 1940. Their job had been to improve the road and build a rail track as this area was planned to become a major supply route up from Suda Bay.

"Saunders, take your men and dig in on 42nd Street." Though many of the men were weak and we were down to less than half strength, I hurriedly proceeded to get the men of the 2/7th into formation and off we marched to join the other units who were already in position. Upon arrival at our destination, the men were instructed to spread out in a nearby olive grove and in the ditches along the sides of the road. This enabled us to have a straight view towards a rise about 50 metres down the road. Our orders were to wait until the Germans came close before attempting a counter-attack so as not to expose our position. We remained silent, ready and waiting for the coming attack.

Earlier in a dawn attack, the Germans had captured a number of British troops. As the Germans pushed forward, the British prisoners were lined up and forced to walk in front of their captors as a defensive shield. The poor buggers didn't stand a chance. That experience was not about to happen again if we could help it. So we waited, tense but patient and fully focused.

It was around mid-morning when the message we had been waiting for was received. The Germans were on the move again and heading our way. Suddenly I saw a German soldier who was obviously scouting the area stand up in clear view. Carefully I took aim and made the shot, it was my first sure kill. The sensation I felt for a moment was the same as I experienced when shooting at a kangaroo out in the bush back home. It was a remote feeling. As I snuck up to the spot where the German lay dead, I made the mistake of looking down at him. He was a blond, blue-eyed bloke probably not much older than me. I felt terribly sorry about shooting him, I wished I could say to him, 'Come on old fellow, get up and let's get on with the bloody game,' you know … thinking football. I dragged him off the road and laid him in a ditch so the men following him wouldn't be alerted to our position.

It wasn't long before the Huns battalion were upon our position and things got very serious and then weird. Suddenly one of the Maori soldiers stood up holding his Bren gun and started doing the Haka! How he wasn't killed is a miracle. The New Zealand 21st Battalion Maori's burst from their cover and all started shouting the spine chilling battle war cry.

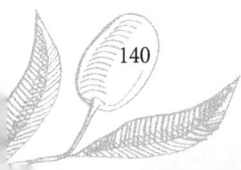

Now we had a new weapon to use against the Germans, our vocal courage. The Australians followed suit and started a vicious scream followed by the Greek soldiers adding their Hellenic yells to the mix. The Germans were stunned by the terrifying noise. Not knowing what had hit them, they turned tail and started running for their lives. We charged and pursued them. By now a mix of all of our units 2/8th, 2/7th, 21st, 28th, 19th, 22nd and 23rd Battalions were on the move forward. We may have been losing the battle overall but just like the heroic Cretan people, we were not defeated. I believed the action undertaken by these men, that day on 42nd Street, would go down as another first in ANZAC military history.

By mid-afternoon the order was given to withdraw from 42nd Street as more German forces could be seen moving around the foothills of the Malaxa escarpment. Staying was not an option. We could not be surrounded. It was time to go and join up with our fellow troops and head for the White Mountains and rally at Sfakia on the south side of the island.

The Germans sustained a high loss of life that day with over 200 of their men killed in the skirmishes. The 2/7th had to bury ten good men of our own but our efforts bought precious time for our fellow troops to escape to the south side of the island. After our charge the Germans were wary and didn't pursue us.

Douglas Channell, Retimo and Hill A

(Douglas Channell)

At the dawn of another beautiful clear and bright morning on 20 May, I had the men perform the usual morning routines followed by breakfast. By mid-morning our interest was aroused by a distant hum and by the sighting of transport planes to the west of our position. As they made landfall they swung further west towards Maleme and Suda Bay. The expected airborne invasion had finally arrived.

The men were warned to keep their eyes open. Not long after midday a further convoy of attack planes flew in and headed towards our position at Heraklion. A mix of Dorniers and Me-110s arrived and systematically bombed the area around the airfield, careful not to land any bombs on the actual airstrip itself. Once they had exhausted their bombs they flew over strafing the area with machine gun fire.

We waited, silent and securely camouflaged part way down the terraced hill, hidden from prying eyes. Our ruse of dummy guns and trenches had worked. Earlier reconnaissance flights carried out by the Germans showed no defensive positions for them to aim at so they were firing in the general vicinity of the airstrip and dummy trenches. The Germans believed they had destroyed the reportedly small force of Greek soldiers guarding the airfield.

Around 1630 hours, we could see another wave of aircraft heading east of Hill A. This time wave after wave of transport planes flew over and started to unload their lethal cargo of paratroops. It was an unbelievable sight. Hundreds of black dots that morphed into a colourful display of silk parachutes started floating down to earth. We soon worked out white parachutes were worn by the troops while the other colours represented cases of stores.

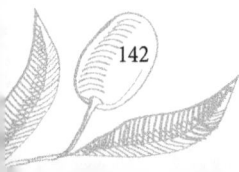

When the paratroopers got within firing range we let them have it. The gunners aimed at the planes, firing short bursts into the pilot's cabin followed by long bursts at the exit doors where the paratroopers were trying to exit. We managed to bring down nine planes and sent many others back to Greece damaged or on fire. Time and again my men performed with discipline and personal courage seeking to achieve their allotted task of defending the airfield. Ill equipped though we may have been, our aim was accurate and deadly. We had a unified spirit, working well together and were ready to give the Germans a run for their money.

Hill A bore the brunt of the attack. Hundreds of the enemy were killed on that first day as they fell from the sky, although I must say one of the men likened the experience to shooting ducks at a sideshow. The paratroopers were totally unprotected as they glided gently to earth and made easy shooting.

Those who did survive the drop and landed safely had to divest themselves of their parachutes, gauntlets and knee pads. Then they had to hunt around for the crates of supplies and form themselves into groups. The Cretans killed many of those who landed. The local villagers used any weapon to hand: pitch forks, knives, spades, sticks and bare hands to kill their enemy. The Germans were totally unprepared for the viciousness and resistance of the local people as they had been assured the inhabitants of the island would welcome them.

When possible, groups of my men were sent out to recover what they could of the crates of stores, which had been dropped by the Germans. We found everything from weapons, ammunition, food and signal gear to medical stores. We were not the only ones searching for these crates, the Cretans also joined us in capturing as much as they could in order to deprive the paratroopers gaining access to these vital supplies.

Those who did land safely, out of range of our gunfire, grouped quickly, located the cases that had been dropped for their benefit, armed themselves and proceeded to come at us in an organised attack. Others, who managed to land almost on top of us quickly took out their small arms and engaged us in a gun fight. Not many of these survived.

The vines and vegetation that protected us also gave these well-trained Germans cover. Once they gained control of the top of Hill A and secured their equipment, they bombarded us with machine gun, mortars, light howitzers, grenades and small arms.

The fighting continued well into the night. Hundreds of Germans continued to group and attack my small garrison of soldiers. By now the Germans held most of the ridge top and eastern slopes. We continued to hold onto the remaining positions on the lower western and southern sectors of Hill A. Unfortunately we had no grenades to retaliate, but we held our position.

At one stage the 2/1st Transport Section of HQ Coy were forced to come around from behind and engage in a charge using bayonets. Our men chased the enemy away from where they had been desperately trying to reach a crate of weapons that had landed close to where 2/1st was positioned.

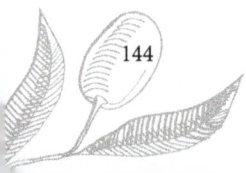

We had no communication with the other units in the area as the Germans had cut the wiring. It was only the bravery of gunners volunteering to act as runners that allowed us to keep in touch and find out how the other units on Crete were managing. Many of us were operating by using captured weapons and ammunition. The parachutists were still being dropped from the sky. One of our machine guns was enveloped by a parachute. The men made quick work of the German and his parachute.

Meanwhile back in Greece at German High Command they were going through an anxious period. Austrian born Major General Ringel realised their assumption that the landing fields were held by small weak Greek forces was obviously not true. He had lost thousands of his highly trained alpine troops and those who had managed to land safely and regroup held their ground only with extreme difficulty and with heavy casualties. At the end of the first day's attack no clear picture of the situation on Crete could be formed.

The Germans around Retimo had made their headquarters in the olive oil factory just outside the town. They put up a very determined bid to capture Hill A. My men in A Company were reinforced by a second platoon from B Company and two from D Company we were able to hold our isolated position half way down the hill. Our wire entanglements stood us in good stead and we were able to keep the enemy at bay.

Our stretcher-bearers carried out a mighty job of retrieving our wounded and tending to their wounds. Their first aid efforts eased much suffering and pain and the men were very grateful for their courage and devotion to the wounded.

Thankfully for us, the Royal Navy were able to intercept an attempted seaborne invasion on the second day of battle resulting in approximately 2000 German troops being drowned or killed.

At 0500 hours I had the men assembled and organised in one extended line. I took centre position and had officers on the left and right flanks. At 0525 hours under the cover of darkness we moved forward at walking pace towards the enemy positions some 140 metres ahead. After covering 90 metres the enemy started firing at us with machine guns. We dropped to the ground and fired back. Some of my men and I were able to continue on and take out a machine gun post.

Next thing I knew I felt a dreadful pain across my back. I had taken fire and was wounded in the back of my shoulder. Realising I was shot up pretty badly I ordered the men to prepare to pull back to our backup support behind us. The men continued to fight and eventually the vital position of Hill A was secured with the surviving Germans escaping to the beach.

My wound extended from above the clavicle to under the middle of my back on the left side, effectively putting me out of action. I was lucky to be alive. After I was taken to the field hospital Captain Embrey took over the defence of Hill A and established a cemetery nearby, where our dead were reverently interned.

NOTE: In 1949 Douglas Channell was awarded the Military Cross. His citation reads:

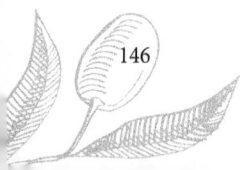

Citation of NX.115. CAPT. D.R. CHANNELL

Conspicuous gallantry & devotion destroying machine gun post.

Capt. D.R. Channell, commanding A. Coy., was responsible for holding a vital Knoll which commanded Retimo Aerodrome and the whole area and coastline in the vicinity. It was bare of trees save for an occasional fig tree, but was completely covered by vineyards which gave excellent cover to snipers and machine guns alike. At 16.00 hours on 20 May, 1941, approximately 150 paratroops were landed on the hill and another 700-800 immediately to the east of it, who, on landing, concentrated on advancing against the Hill.

Fierce fighting ensued and continued the remainder of the afternoon and night, and by 22.00 hours of the night 20/21st May, 1941 the enemy had gained control of a large portion of the Hill and would undoubtedly have advanced further but for Capt. Channell's handling of the situation.

Capt. Channell was ordered to attack at dawn next morning to clear the Hill, additional troops being placed under his command and 600 Greeks were ordered to attack on his right flank. Although the 600 Greek troops did not materialise, Capt. Channell personally led the attack against a numerically superior enemy and over country ideally suited for defence.

He was badly wounded in the back shortly after moving forward, but continued to lead his men with conspicuous gallantry against heavy machine gun, mortar and rifle fire.

He preceded his men up to the abandoned gun positions, encouraging them the while, finally destroying a German machine gun post with grenades, thus very materially assisting in the advance.

As the only medical unit in this area it was impossible for the 2/7th Field Ambulance to evacuate wounded after the battle started. Because of our lack of communication with units on the rest of the island it was unknown to us at Retimo but the entire British Force including 2/4th Australian Infantry Battalion defending Heraklion were evacuated by sea on the night of 28 May. Apparently three separate messages were dispatched to our force at Retimo ordering our withdrawal to a small inlet on the south coast of the island.

As these dispatches were never received, our orders still held. We had to hold the airstrip against air and sea landings. It was our duty to stay. With the German infantry battalions, tanks and light artillery pieces advancing on our position, and only enough rations for another day, our CO Lt. Col. Campbell had little choice but to make the painful decision to surrender. Given the choice before capture some of the men opted to head for the hills and make their escape. Some of these men eventually made their way to Africa.

Once the Germans took control of the airfields they were able to land their gliders and disgorge hundreds of ground troops. The Allies who remained were officially declared prisoners of war. After the officers had been interrogated all of the men who could walk were marched away to Suda Bay.

Around eighty of us were left at the little hospital at Ádele. It was here where I too was captured and taken as a POW and eventually sent to a POW camp in Austria.

Years later I was to learn young Marcos, our runner, developed a searing hatred for the Germans. The day 'the Germans fell out of the sky on silk clouds followed by silver ghosts' marked the start of the destruction of his family.

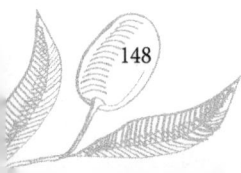

Reg Saunders crosses the White Mountains

(Reg Saunders)

As the Allies retreated, the 8th Greek Regiment, a force of young recruits, gendarmes and cadets, held the Nazis at bay. For seven days they fought with legendary courage and tenacity. All of them were either taken prisoner or killed. They were true heroes. In their villages, civilians armed with makeshift weapons, were ready to fight and defend their island in any way they could. In the future they would pay a heavy price for defying the Germans.

Hitler had hoped for a swift victory but this was not to be. He could not believe the small island of Crete could hold out longer than the whole of France. He was forced to send more and more troops to Crete in order to gain control of the island. Not an ideal situation for him and this unforeseen hurdle further delayed his plans to invade Russia. Instead of the expected three-day siege, the battle raged for twelve days.

With the battle lost the Allied forces staged another co-ordinated withdrawal and headed across the White Mountains to Sfakia. In full retreat and while we still had time the 2/7th and the 2/2nd fought some challenging rear-guard battles to further delay the advancing German forces.

I was so pleased to reach the cliffs of Sfakia. It had taken us five days to complete the tortuous retreat through the White Mountains with the Germans not far behind us. We climbed down the cliffs, tired, hungry, thirsty and dispirited, only to see the ship full to capacity and getting ready to set sail. With the thousands of men on the beach we knew there was little hope of another ship arriving before the Germans did.

Colonel Walker was standing at the back of the last barge ready to escape when he turned and saw us, his men, who were a part of the contingent being left behind. He jumped into the water and joined us on the beaches.

Walker addressed me and the other members of the 2/7th, telling us he was going to surrender, however if anyone wanted to take their chances and escape this would be their last opportunity to do so. Though we were bone tired from the five-day march and for the most part out of ammo, a few of us said our farewells and hightailed it off into the night.

The official surrender of Allied forces on Crete was reluctantly given by Colonel Theo Walker, commander of 2/7th Battalion to an Austrian officer of the 100th Mountain Regiment at Komitades, near Chora Sfakion, on 1 June 1941. Colonel Walker was among those taken prisoner.

The Royal Navy successfully evacuated over 12,000 men from Sfakia, regrettably more than 5,000 Allied troops were unable to make the evacuation. Most became prisoners of war, some were killed and a small number managed to make their way back to Egypt with the help of the local people.

Following is an excerpt from the poem, The Sixth Division Saga – 1939 to 1945 written by QX4567 Ex. Sgt. T.J. Kemp 2/1st Tank Attack Regt 6 Div. A.I.F.

' ... *Some ships went back to Alex, many others went to Crete,*
Where the ANZACs and the British, air invasion was to meet.
Though ill equipped and poorly fed, they fought the German flood,
Of parachute and glider troops – our brave men gave their blood!

In spite of heavy losses, the fresh Germans won the day,
They poured in reinforcements on those fateful days in May.
Our weary troops gave all they had to stem the aerial tide,
But once again it was withdrawal – to the waiting Navy's side.

So all was well for those who lived to fight another day...',

The heroic actions undertaken by the 8th Greek Regiment saved the lives of so many Allied troops, Winston Churchill was inspired to publicly state:

"Hence,

we shall not say that Greeks fight like heroes,

but that heroes fight like Greeks."

CHAPTER 8

Death at Suda Bay

Is this for real!

(Dimitri James (Jim) Zampelis)

"Hey Jim, how about you share your story with us tonight!"

"Ok," I replied. Sitting around the table while stationed in the Middle East with the guys having a few drinks made the telling of my story a pretty informal affair.

My Zampelis ancestors hailed from the village of Marantohori on the Island of Lefkada, off the western coast of Greece. In 1900 my father, Gerasimos, made the ambitious and adventurous decision to leave his family and homeland to embark on the long sea journey to Australia. Dad eventually settled into the small Hellenic community of Melbourne in 1903.

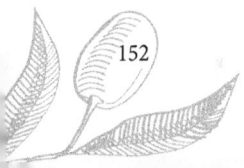

It was there in 1910 that he met and married my mum, Louisa Elizabeth. My sister Helena was born in 1911, myself, Dimitri James, in 1912. I have always been known by my shortened second name Jim. I was only seven when Mum died.

Dad and his cousin Nick were the proud owners of Nick's Café. As a youngster it was only natural to find me in the café after school and on the weekends working as a waiter and washing dishes.

"Not much has changed, hey Jim! You're still washing dishes today," called out Tommy. We all had a good laugh at this comment.

I told them how I married Doris and we had a baby boy, Peter. Unfortunately the marriage broke down and Doris and I separated. In mid-1939 I was enjoying life, working, spending time with my young son, family and friends. I had a good job working as a waiter in the refectory and living at the prestigious Newman College in Melbourne. It was good fun and I liked working there. When the war broke out in September, life was to take a significant turn for me. Within four short weeks I had put my affairs in order and enlisted.

Jim & family

By the time I finished my story it was time for lights out. I fell straight into a deep sleep but started dreaming about what had happened since I joined up.

Separated from my wife Doris, I nominated my young son Peter Jim Zampelis as my next of kin on the enlistment papers. It was hard leaving my son but I knew his mother would take very good care of him. My father had also promised to keep an eye on him and send me photos of Peter with his own letters. Funny how something as small as a photo could mean so much when you were going away for goodness knows how long.

Considering my background in hospitality, it came as no surprise they assigned me a job as mess steward in the newly formed 2/2nd Australian Field Regiment. The 2/2nd took part in a number of major engagements in the Middle East.

For some months things were pretty hectic in the Middle East and my letters to Dad had been short. I remember taking the opportunity one afternoon when things had settled down to write a long letter to Dad to let him know what had been happening with me.

Wednesday 11th September 1940

Dear Dad,

Having received your telegram on 14th August I am still waiting to receive one letter.

I suppose you have got my Air Mail letter in reply to the telegram.

We have now arrived in Egypt and everyone thinks that it is 100 per cent on Palestine. There is as many interesting things & places to see, as we can cope with in ones spare time, we are not hard done by in work, but, at the same time, we are not hard done by in the way of leave. I have had a look at Cairo, & what a place it is.

Apart from the popular places that, for years have catered for the tourist trade; there are at least five clubs established especially for the Army. One place, called, "Empire Services Club" is a regular cabaret.

A good 7 piece band is playing every evening, with floor shows every few minutes. The food is excellent, & the prices are very reasonable. There are (apparently) no licensing laws to worry about so we can always get a bottle of ale with meals if we wish.

When the boys get back to Victoria again the six oclock closing is going to be hard to get used to again.

I paid a visit to a small town not very many miles from Cairo, called Helwan, & discovered some lovely gardens there. You have no idea how good it was to just sit there with the green lawns, trees & flower beds all round. If I get back, I'll never take Melbourne gardens as a matter of course again. It takes a few months in a semi-barren country like we have been camped in, to make us realise just what it is to be able to enjoy the quiet & coolness of the gardens.

We are hearing all sorts of rumours, but taking no notice of them. When things happen, we'll be well in it, so we are just waiting for them to happen.

How is Peter getting on, Dad, don't forget that photo of him. I hope Doris, Nick, Harold, Len, & yourself are all in the best. I am & hope to remain so. Tell Harold I see a lot of Arthur Longhton [sic] an old cobber of his, & he wishes to be remembered.

He wants to know, does Harold remember those dinners they used to have at the flat? Tell him, that Arthur Hilton from the hamburger shop in Acland St is here also. (Arthur used to go to school with us.)

Some more of my old pals from the drill hall joined us just before we sailed. Don't forget to get Harold to send me a copy of "Man" every month, & if possible get me the back numbers from last May. I have some-one sending "Smiths Weekly." And some-one, whom I don't know, is sending the "Bulletin" & "Truth."

I have written several letters to Len, I hope she got them all. I don't know just how many there is. If you people get together, see if you can arrange for one or other of you to write to me each week, by Air Mail. Letters are more than welcome & as far as possible will be answered. We have plenty of time to write now, but it may be different later, so, if they get fewer, don't get disgusted & give up writing.

You can see by the crest where this paper comes from & where I am writing, so, Dad, when the, "Lassie" comes in again (on Fridays wasn't it?) give her a extra. They do a lot towards making things comfortable for us.

Well, I've run out of news now so will close. Wishing you the best of everything.

I remain

Your loving son

Jim XXX

P.S. Remember that steak I wrote about in my last letter? I got my teeth in to one, in Cairo, & what a beauty.

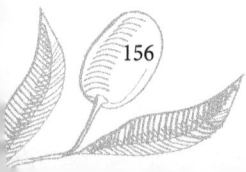

Delivering good meals for the officers and providing canteen services for members of the unit was vital to the well-being of our deployed personnel. Meal times were when the men received their tucker, and got to relax and unwind.

My role of mess steward consisted of many duties, including stocking food, cleaning and assisting with the preparation and serving of meals. I was also responsible for seeing to general stocks such as food, linen and utensils and making sure they arrived where they were needed. I was closely involved with the storeroom operations, particularly when it came time to take inventory. All of our linen supplies needed to be sorted, counted, stored and issued as directed.

I enjoyed assisting the chief cook with the preparation and serving of the meals. For me it was just like working for my father or the head chief at the college. Most of these tasks were second nature to me, they were things I had been doing my entire life, though I admit under totally different circumstances.

My unit had been involved in the fighting in the Battle of Bardia. Code-named Operation Compass. This battle took place between 3 and 5 January 1941 and was where the first battle fought by Australian troops in the Second World War happened. The town of Bardia was captured in the late afternoon of 4 January but the Italian resistance in the southern area continued until they were beaten and capitulated. It cost us dearly with over 100 dead and more than 300 wounded. Around 40,000 Italians were taken as prisoners of war, which meant more mouths to feed. Thankfully, our lads had been able to salvage massive quantities of arms, rations, equipment and alcohol. Believe me when I say we Australians put this to very good use.

We were starved for news from home. Letters from friends and family were precious and mail time was a much-anticipated event. Books, magazines and old newspapers were also much sort after items that could be shared among us. While this was a very important time in our lives, and we had volunteered to go and fight to keep our Allies and Australia free of German tyranny, it was also a very lonely time and hard for us to be away from our families for such an unknown period of time.

While the unit was having a break from fighting in early 1941 I found time to write the following letter to one of my mates from Melbourne who had also enlisted:

Monday 3rd Feb 1941.

VX929 Gnr J Zampolis
R. H. Q
2/2 Field Regiment
A.I.F. ABROAD.

Dear Jack,

Len has mentioned several times in her letters that you have come over, but it was not until I got her last letters (today) that she managed to include your address.

How do you like Palestine, Jack. It's not quite so good as Aussie land is it. But it's not so bad. We have managed to get mixed in with the Italians once or twice, but they are not so hot. Don't like to play, in fact.

How is Tel Aviv & Jerusalem looking, now? You had better be around Cairo or Alex by the time we get back. We have a date, don't forget. Do you ever hear of Lofty? He stayed back when we left, & I've not heard of him since.

Heck! its cold tonight, Jack, and raining too. Feeling a shiver now, so don't know how it will be when I get to bed.

I suppose your blokes have to dig slit trenches, we will & but here we sleep in them as well. It's not so good when you wake in a few inches of water. A week or so ago we slept in a house for several nights & the second night the plumbing went wrong. Result was wet sails for my oath & myself. In a house too, above all places.

Well, Jack, don't forget to answer. Solong.

Jim.

P.S. When I wrote to Joyce I told her you came over she answered that the Italians ought to hear Joan to say N-n-n-no & you'd just fade away. Remember that night!

A. Zampolis

Monday 3rd Feb 1941

Dear Jack

Len has mentioned several times in her letters that you have come over, but it was not until I got her last letter, (today) that she managed to include your address.

How did you like Palestine, Jack. It's not quite so good as Aussie land is it. But it's not so bad. We have managed to get mixed in with the Italians once or twice, but they are not so hot. Don't like to play, in fact.

How is Tel-Aviv & Jerusalem looking now? You had better be around Cairo or Alex by the time we get back. We have a date, don't forget. Do you ever hear of Lofty? He stayed back when we left, & I've not heard of him since.

Heck! It's cold tonight, Jack and raining too. I'm doing a shiver now, so don't know how it will be when I get to bed.

I suppose you blokes have to dig slit trenches, as we did, but here we sleep in them as well. It's not so good when you wake in a few inches of water.

A week or so ago we slept in a house for several nights & the second night the plumbing went wrong. Result was wet tails for my cobber's & myself. In a house too, above all places.

Well, Jack, don't forget to answer. So long.

Jim

P.S. When I wrote to Joyce & told her you came over she answered that the Italians ought to hire Joan to say N-n-n-oo & you'd just fade away. Remember that night?

A month after I wrote to Jack we received word to say we were deploying to Greece. I was very excited by the news, of leaving Africa to journey to the land of my ancestors. Bring it on!

When I arrived in Greece in April 1941, it was like walking into a fairy tale. Words could not express the great pride I experienced as my feet first touched Hellenic soil and I walked down the roads my descendants had travelled for hundreds of years. All the places my father had told me about over the years were right in front of my eyes. The warm-hearted people, the lush landscape, the crops, vines and fields of wildflowers, magnificent buildings and monuments; Greece the home of myths and legends lay spread out before my very eyes.

It was almost like having an out-of-body experience. Excitement welled up in my heart so much it almost overwhelmed me. To actually be here, standing in the land of my ancestors for the first time in my life. I was extremely excited and had so much to write home and tell Dad about. The feeling of coming home was powerful, I really felt at peace here.

Knowing the language gave me a distinct advantage over others in my unit and resulted in me securing some very good bargains. Needless to say I had to do a fair bit of translating for my mates and the officers. It was all good fun and filled in time during those first few days, good people, good food and great sights to explore.

One thing I enjoyed was getting an opportunity to shop at the markets and farms close to our base to secure fresh fruit and vegetables for the mess. These were foods I was familiar with and knew how to cook. The head chef had me working closely with him in the kitchen, teaching the cooks new dishes using this fresh Mediterranean produce. Fresh olive oil and plump ready to eat olives, life was good.

Many of my friends had never tasted an olive before, the taste was not something I could describe to them, they needed to try them and try them they did. They loved the fresh briny flavour and they were great with a beer or glass of wine. I also introduced them to a personal favourite of mine, fresh bread dipped in olive oil.

I was comfortable with my job in the mess and the officers were always pleasant, appreciating how hard we in the kitchen worked in our endeavour to keep everyone fed and healthy. It was their duty to keep the men safe in the field and ours to keep them fed and healthy.

This was difficult when we were constantly on the move. Sometimes the mess tents were hoisted in some rough and ready areas, but our job was always to make sure everyone got fed well and didn't get sick from the food.

At least in Greece we didn't have the problem of hordes of flies, sand in everything or the heat we had experienced in the Middle East. It was the exact opposite here. High altitudes, rugged mountains and snow. Life can be tough in the army, particularly when you are on the front line in the thick of things.

A good old M and M stew from the mess tasted like heaven compared to the bland, though sustaining, ration packs the men sometimes had to revert to eating. At least the C-Pack keeps body and soul together when needed.

Sometimes the men would sing a ditty about the cooks, which went something like this:

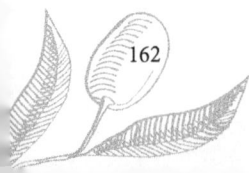

We may brag about deeds well done;
On battlefields we have won;
But let's not forget one special group;
To whom we shout "where's the gloop";
The one's who cook the mountains of stew;
And follow up with a darn fine brew;
Their not greasy, grimy chooks;
Their our flaming great cooks.

It was only a few days after we landed in Greece that we received orders to head north. Many units formed together in the mountains, desperately digging in using every effort to keep the Axis at bay. It was here at Vevi Pass where fellow Australians, Douglas Channell, Alfred Carpenter, Ronald Collins, Alan Eason, Clifford Morris, Albert Mayer, Reginald Saunders and Glen Scott and their respective units all took up positions to defend Greece.

In a war zone you need to be adaptable, regardless of your normal role. If I was needed to make beds in the officer's quarters, I made beds. If I was needed to work in the infirmary kitchen, I worked there. If I was rostered onto canteen service duty, that's where you would find me. If it needed to be done you just did it, no questions asked. These men were your 'family' and you did whatever it took to cover each other's back and make life easier all around.

Constantly keeping a supply of boiling water so those taking a rest break could have a good strong hot cuppa, particularly in the wretchedly freezing cold mountains at the Vevi Pass is one example of how such a tiny thing we would take for granted at home could make such a difference to someone's day on the war front.

The guns of the 2/2nd started their campaign in Greece on 16 April at the Vevi Pass. When ordered to pull back, the unit engaged in short, dogged defensive actions, followed by some clever and strategic withdrawals through the mountains and valleys. Feeding the men on the front line is not an easy task, feeding them on the run is almost impossible. Trying to get food for them and making sure they had ration packs became my main focus.

At Lamia, I realised I was directly east of the Island of Lefkada, the home of my father and his ancestors. The island was approximately 250 kilometres to the left of our position. In different circumstances it would have been nice to have had the opportunity to travel to Lefkada and walk the roads as my father had, but this was not to be.

From 21 to 24 April, the guns of the 2/2nd held up the German advance across the Spercheios River. Our gunners were subjected to sustained German artillery and aerial attacks – in one two-hour period we suffered well over 100 aerial bombing attacks by scores of dive-bombers. This was followed by a further eight hours of enemy artillery rounds. It felt like my ears would never stop ringing from the sounds of battle.

Our good camouflage and subterfuge meant casualties were slight, with only five killed and three wounded. As cooking was now out of the question it was a matter of helping the cook keep rations up to the men and supporting the medicos with the wounded. We took care to bury our dead, mark their locations and record all their details. Thankfully there was only a handful, but every man buried was one too many.

After this rear-guard action was completed, we withdrew south over the mountains to avoid the German air attacks, travelling on temporary roads built earlier by Australian engineers. Riding in one of the trucks, I would jump out to help wounded soldiers who couldn't put one foot in front of the other. I would get them safely into the back of the truck and then we would be off again. This continued for as long as I could manage to fit them in. Our arrival at Megara happened to fall on Anzac Day 1941. After spending the day resting under olive trees, it was time for the unit to embark on Allied sea transports headed for Crete.

Despite having dragged their guns and ammunition over the length of Greece, the 2/2nd gunners were forced to destroy their armaments to make room for evacuating troops. Like so many other Allied units, mine would face the next battle without their heavy artillery. For now we were issued with rifles and what little ammunition was available. A far cry from the familiar Bren guns and heavy artillery my unit were trained to use.

I gathered what few rations were left in the mess supplies and made every effort to load up each of the men with some food. The remaining cooking equipment was given to some of the local Greek people who had been helping us. Other items were rendered useless. Nothing useful was left undamaged thus preventing the Germans from gaining any spoils of war. We sailed all night and arrived in Crete the next day. We were instructed to bunker down at Suda Bay.

From dream to reality

(Dimitri James (Jim) Zampelis)

Reveille sounded and I woke up with a start. What was going on? I realised I had just been dreaming about my life. "Wow, what a dream." I shook my head and got ready to face a new day.

Crete was a very picturesque island. It was wonderful to meet the local population and speak freely in Greek to them. Once again my interpreting skills came in handy for my superiors and friends. These people welcomed me like a long lost son. Even though the majority of the population were as poor as church mice, the hospitality took me back to memories of my father in Melbourne. Nothing was too much trouble and the best you had available was always given to your guests, even if it meant you went without. This was my heritage, something I had grown up with, even though I was born and lived on the other side of the world. My blood was full of this heritage and these traditions. I felt truly at home here and fell in love with the beauty of Suda Bay despite all the war, troops and supply ships harboured there.

Perhaps it was because the port of Suda Bay was such an ideal harbour that Commanding Officer, Major-General Bernard Freyberg believed Germany would predominantly strike there by sea and not by planes at the airfields on Crete, which were all under Allied control.

From 4 May 1941, the 2/2nd were assigned the task of helping the Australian engineers unload the ships. This was done with great pragmatism with the men only stopping during the bombing raids. A number of highly successful salvaging expeditions were carried out during this time, locating and retrieving artillery, ammunitions and anything of use from ships that had been sunk in the harbour.

The Germans launched Operation Mercury on the morning of 20 May 1941. We couldn't believe what we were seeing in the sky. Hundreds upon hundreds of planes dropping thousands of paratroopers.

Orders were given to move and we found ourselves stationed in the village of Mournies. The Germans had been attacking the village for a number of days but had been repelled each time. Four days after the start of the attack, in the afternoon of 24 May 1941, I was assigned the duty of helping the wounded men at a sick parade in the village. Unfortunately, the village was dive-bombed. I could see and hear the screaming of the planes as they descended from the heavens. Looking up I could tell the strafing would be close...

Dear Mr Zampelis

The following letter written by Lieutenant Colonel W.E. Cremor, Commanding Officer, 2/2 Australian Field Regiment, was received by Jim's father in July 1941.

> *July 1st 1941*
>
> *Reg. Head quarters*
>
> *Dear Mr Zampelis,*
>
> *I am writing to convey to you the sympathy of myself and members of my Regiment on Jim's death. Jim was known very well to me personally, as he had been with Regimental Head Quarters since the regiment was formed. He was always cheerful and reliable and did his job courageously. He was popular with us all because of his pleasant disposition, and we feel his loss intensely.*

On the afternoon of May 24th, when the Regiment was in action at Canes, (Crete) the Germans launched a fierce bombing attack, which killed Jim and several lads. They were all killed instantaneously, and that evening, they were buried at Canes by men of the Regiment.

Again assuring you of our very deep sympathy.

Yours sincerely

W. Cremor Lt. Col.

Commanding Officer.

2/2 Aust. Field Reg.

The Battle of Crete raged for ten days from the morning of 20 May until 1 June 1941. While this was to be Jim's final resting place, his family could be justly proud of the spirited resistance Jim and the Allied troops put up against the invading forces of Germany. The people of Jim's ancestral homeland fought so courageously alongside the ANZACs of his own homeland, Australia. Jim's service records indicate he was buried by his comrades "500 yards south-west of Mournies village".

Sadly after the war, the Commonwealth War Graves Commission was unable to locate Jim's body, so he never received a formal burial. Jim, together with other brave soldiers whose bodies were never recovered, are memorialized on the Remembrance Walls at the Phaleron Military Cemetery in Athens. The remainder of his mates, whose bodies were recovered, now lie at rest in the Suda Bay War Cemetery. Hopefully one day in the future he may lie at rest beside them.

CHAPTER 9

The Resistance and Cretan Civilians

A young Cretan youth

(Timothy (Tim) Lionakis)

I am Timothy (Tim) Lionakis. My father migrated from Crete to America sometime in 1909. This is where he met and married my mother. I was born on 26 December 1924 in Philadelphia. By 1929, the Great Depression had brought America to its knees and life was very difficult for my family.

Life took an even harder turn when my mother died leaving my younger sister and I in the care of our father. The law at the time stipulated a single father would be unable to care for the children properly, requiring my father to hand us kids over to the government to be put into an institution.

My father was heartbroken at having lost his beautiful wife and he refused to relinquish us, his beloved little children, to the government. Dad hid us until he could make plans to smuggle us by boat back to his family in Crete. A kind family offered to claim us as theirs and we sailed to Greece and then onto Crete with them sometime in 1931.

I loved my life on Crete. It was a simple life, which offered a small boy a lot of outdoor adventures. It was a simple but productive life. The people were poor and when I first arrived at school wearing socks everyone thought my family was rich. This was not true, it merely meant in America we wore socks with our shoes and my father had packed them for me when he sent us to Crete. I was a quick learner but decided school was not for me so I left after only completing one full year.

I learned many valuable skills. It turned out I had a natural flare for languages and very easily learned the local dialects. For many hours I would wander the rough but picturesque landscape of the island, learning the geography of my new home. My grandfather taught me to shoot and hunt rabbits enabling me to supplement the family's meals. I also enjoyed fishing, trapping, looking after the goats, growing vegetables and tending our snail farm.

Grandfather

On the morning of 20 May 1941, I was sitting at the kitchen table having breakfast oblivious that this day would see the end of my childhood. Life for everyone on the island was about to be torn apart and there was nothing we could do to stop it happening.

My grandfather rushed into the house in a very agitated state. He was in his early sixties and this behaviour was very out of character for him, which made us realise something was very wrong.

Grandfather gathered the family together and told us what he had seen and explained why it had so upset him. While on his usual morning walk up the hillside to tend to his small herd of goats, he heard a strange humming sound. He stood still and listened carefully. This was a sound like no other he had heard. The humming sound grew louder and louder. Suddenly in the distance he could see dozens upon dozens of planes flying straight for the island. The threatened invasion by Germany had arrived.

He watched in horror as the early planes bombarded the area in the distance around Suda Bay. When they stopped and turned back to Greece, more planes took their place. Plane after plane crossed the sky dropping thousands of paratroopers onto the island. These men were floating down to earth as he ran back to the house to warn the family and to arm himself.

He was furious! Crete had been declared a free island in 1913. After 400 years of occupation by the Ottoman Empire and others, they had won their independence and were a free people. Now, like his ancestors, he would be forced to pick up a weapon to defend his beloved family and island home.

Vicious, disrespectful Germans

The Cretan people were involved in the Battle of Crete right from the beginning. Any paratrooper unlucky enough to land and come into their range was killed or taken prisoner with fierce brutality.

These elite German soldiers had been briefed by their superiors and told the local population would offer no resistance. Their shock at being brutalised and seeing their fellow paratroopers killed was incomprehensible. They had never experienced this type of resistance from civilians and it led to many men falling victim to a people determined not to be conquered again.

Even when the German command and their soldiers on Crete had beaten the Allies, they continued to underestimate the defiance and resilience of the Cretan people and their loyalty to the ANZACs in particular. This loyalty to the Allies came at a heavy cost to the Cretans. My people were to pay dearly for helping the Allied soldiers, but we did not hesitate or stop to consider the cost.

The Cretans had embraced the Australians and New Zealanders. We acknowledged the fact these men understood and honoured our traditions and beliefs. These soldiers from half way around the world had respected the high values our Cretan families held regarding their women and they shared many of the same beliefs. The villagers would not abandon our new friends to our common enemy.

The Cretans involvement in killing of German soldiers led to brutal reprisals and counterinsurgency methods being put into place by the Germans. None were safe; men, women and children were taken hostage and many were brutally executed. The Germans were furious with the people who refused to be cowed by the mighty Third Reich.

Thousands of Australian, New Zealand and British troops had little hope of escape from the island now the evacuations had stopped. German soldiers and airplanes patrolled endlessly trying to locate and take these men prisoner.

Tragically, when the last of the soldiers capitulated at Sfakia the Germans were not able to advise the Luftwaffe in time to stop any further attacks. As some of the allied soldiers stood on the cliffs above Sfakia the Luftwaffe flew over with machine guns blazing and killed many of them.

To harbour an Allied soldier meant death to the families who hid them. Shocking instances took place where entire families—men, women and children—were murdered for defying this German decree. Despite suffering heavy causalities the Cretan people in villages all over the island continued to help the Allies.

It was very difficult for Australian and other Allied soldiers left stranded on Crete to turn to the locals for help as reprisals were swift and often deadly. The remaining men were never to forget the great debt they owed the Cretans.

The Germans could never hope to win the Cretans over. Their total lack of respect for these people was evident. Unlike the Allied soldiers who respected the Cretans the Germans wanted only to dominate them. They cared nothing for their traditions or values. They took what they wanted by force and if anyone objected they were murdered. This included the rape of the Cretan women. In one of the villages a father shot and killed his daughter after she had been raped and fallen pregnant to a German soldier. She had been so dishonoured the father couldn't live with the thought of what had happened to her.

Manhood

'Tim, you are a fit young man and I need you to listen to me carefully,' said Grandfather. As he spoke I could clearly hear the passion in his voice and see the dangerous glint in his eyes. The time had come for me to stand with the resistance and join Grandfather in the fight to protect our home. Up until this time I had used my old rifle to hunt rabbits for food but now circumstances decreed it be used for a totally different purpose. I needed to be prepared to shoot and take the lives of the invading German soldiers.

I felt proud to join with the others and fight with the resistance; I was a man from Crete. I grabbed my rifle and followed Grandfather out the door. We were off to fight and kill the enemy. By this time everyone in the village had been alerted to the German invasion. Villagers grabbed whatever weapons they could to defend themselves and we all went our separate ways.

The first two days were hard. I had to come to terms with the current situation quickly. I was being thrust into the middle of a war, killing other human beings and constantly looking out for my own safety. I very quickly realised I would do whatever it took to keep my family and friends safe.

The villagers watched on in fascinated horror as the planes overhead spewed out hundreds of different coloured parachutes with German soldiers attached to them. In one of the villages the padre and his son broke into the local museum and without a second thought, confiscated a gun and some ammunition. They tramped out into the field and started shooting the German paratroopers as they glided down to earth. Members of the resistance later told me the son described it like shooting ducks in a sky filled with brightly coloured silk clouds.

Those of us who had rifles took to the areas away from the Allied soldiers' positions. Grandfather and I knew our allies would continue to defend our village, so we decided to search further afield and attack the Germans who were looking to land out of range of the soldiers' gunfire. Once in position we took aim and shot the German paratroopers as they floated down. Those who managed to land safely were set upon and killed where they landed. Men and women fought side-by-side using pitchforks, hoes, large sticks, rocks and anything that would inflict injury or death to our enemy.

Over the next few days as more and more Germans managed to land safely, I decided to go off into the hills on my own. I found a good hiding spot in the brush on the mountainous terrain above the roads being used by the Germans. This position was not within range of the village and was a couple of miles from home. I deliberately chose this spot as I did not want any reprisals being connected to my family or village, plus, it offered me a good escape route.

Sometimes I would lie for hours in the one spot, just waiting for a small group of Germans to come around the bend in the road and into my line of sight. From my vantage point I would take my time, breathe in and out slowly and shoot to kill. Already an accomplished shot, this experience only strengthened my resolve to become an even better marksman.

At times the Germans would round up local villagers as work crews and take them to sites in the area. Often these jobs of repairing roads and other infrastructure would last weeks at a time. Other times the crews would be taken some distance away, making the return trip to the village each evening impossible and they were required to camp where they worked. When this happened they would be guarded day and night by the German guards.

The resistance developed a plan. At night after the men had eaten and were getting ready to settle down for the night, someone would sneak out of the camp and be replaced by another person. When a head count was done the numbers were the same every day.

This was done to exchange information in and out of the work camps so the resistance would know what was happening. With this vital information we could plan our missions to hinder the German's efforts.

One day a light plane flew in from North Africa with General Rommel aboard. I was standing in an inspection line no more than twenty feet away from this man. If circumstances had been different, I would have shot him dead. This man, known as the Desert Fox was hated by all of us.

Angelo Bournazos was a very good friend and mentor of mine. Though he was older, I had much respect for this man. He taught me many skills, including how to handle a knife and use it effectively if I should find myself in a face-to-face encounter with an unarmed German soldier. He also taught me patience and concentration: how to focus and not be distracted and how to use my hearing to the greatest advantage. I learnt to recognise the familiar sounds around me so I could quickly be alerted to unusual, suspect and dangerous noises.

Tim and Angelo

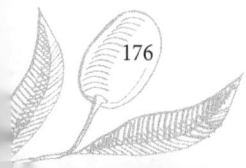

Angelo and I would often go on missions together to disrupt German plans, all the while searching for Allied soldiers who needed our help. Rocky mountain ridges and caves were to become home for a good number of these stranded soldiers who managed, with our help, to elude capture.

One of my duties with the resistance was to set up snail farms for the Australian and New Zealand soldiers hiding in the caves. This required me to collect edible land snails and teach the soldiers how to carefully look after the snails so they would multiply, thus providing them with a constant source of protein. I also showed them the best way to set snares to catch rabbits.

Sometimes a few of the old women in the village would take their baskets out looking to collect wild herbs and lemons from the bush trees away from the village. Hidden in their baskets would be food for the soldiers. This was dropped at a pre-arranged spot, so in the evenings under the cover of darkness one of the men could make his way down from the cave and collect the hidden food. Even the threat of death would not deter these women from helping. I was very proud to be a Cretan.

Marcos

(Marcos Polioudakis)

I was worried when a badly wounded Captain Channell was taken to the field hospital. The Captain's injuries were very serious and I was worried he would die. It was not long after this that the Germans gained control of the airfield and started to round up the Allied soldiers. I was devastated to see Captain Channell and the soldiers at Retimo being captured. These men had been so good to me and I was not able to stop the dreaded Germans and save them. I returned home with a very heavy heart, little realizing what was about to happen to my own family.

I was to lose three members of my family in quick succession. The Germans came and took my father, executing him on 1 June 1941. Two days later they returned and my grandmother was killed because she argued with a paratrooper. My grandfather tried to intervene to save his wife but he was to suffer the same brutal death. They shot them both right before my eyes. In the blink of an eye those people closest to me were gone.

My anger and grief were all consuming. From that day on I swore I would do everything I could to save Australian soldiers from being captured by the Germans. I would search for, hide, feed and look after the Australians who had been left behind on the island. This work gave me a purpose and I helped many ANZAC soldiers escape the Germans.

Reg Saunders on the run

(Reg Saunders)

After Major Walker gave us the ok to scram to avoid capture, a small group of us decided we would make a run for it. Already hungry and tired we had no choice but to make our way back up the steep cliffs as quickly as we could manage and get back to the track we had followed to get here.

It was hard going, trying to navigate our way past the broken down and abandoned vehicles along the narrow track. It was even harder to walk by the bodies of our men who had been too weak and ill to make their escape; men who had laid down and died where they took their last step.

I found a pair of binoculars lying on the ground beside one of the broken down vehicles and picked them up for future use. We knew the Germans would be coming towards us so we opted to zig zag our way down the steep side of the mountains away from the track. It was a dangerous choice but the only one we could make under the circumstances. If our escape was going to work we needed to get as far away from Sphakia as quickly as possible.

We had been walking for hours and we could see first light just starting to dawn when one of the men with me, John, took a terrible tumble down a steep embankment. I told the others to continue on while I scrambled down the slope after him. He had come a cropper. There was blood all over his face from a broken nose, cuts to his legs and he had a badly sprained ankle. Not game to take his shoe off, I went searching for a dead branch in the scrub. After much difficulty, I finally got him on his feet on the improvised crutch under his arm and we continued on as best we could. There was no way we could catch up with the others so we knew we would be on our own. John begged me to leave him and save myself, but I told him we were in this together. Daylight made it vital to find a good hiding place. I found a small stream running through a patch of scrub and we collapsed in a heap. Covering ourselves with dead fallen leaves, safely hidden from the prying eyes of the Luftwaffe, we fell into a deep exhausted sleep, neither of us stirring for twelve hours.

By the time John woke up, I had already done a quick reconnaissance around our location and discovered a goat track heading in a north-easterly direction not far from our hiding spot. I had also been able to snaffle some oranges, figs and wild herbs from a small clearing. Most importantly I had not laid eyes on any German soldiers. When I examined John's leg it was a bit of a mess, and his ankle was very swollen. There was no way I could have taken his boot off even if I wanted to. After bathing his wounds as best as I could, we sat eating the oranges. I don't believe I had ever tasted such sweet oranges.

We waited patiently for dusk to settle into night. There was no way John could walk on his own so when he couldn't walk any further with the temporary crutch, I carried him on my back. Thankfully he was of slight build which made the task a little easier. The going was very slow through the darkness and by the first rays of daylight, after many stops along the way, we rested under an olive tree. Extra care was required while travelling at night as the way was strewn with loose rocks and most of our travel had been uphill. I estimated we had only covered around nine miles and I worried that if any German soldiers were scouting for wandering Allied soldiers they would soon catch up with us.

I kept these thoughts to myself. The next morning we camped under the big old olive trees. The surrounding trees afforded us a lot of protection from the sun. I had not heard a plane since the day before and was grateful for that. John was in considerable pain now. He was running a slight temperature and his injuries were more serious than first thought. Sitting on the grass we shared our last orange before settling down to sleep. Once again I fell into an exhausted but troubled sleep.

During our daylight sleep, John made a decision that was very hard for both of us. Knowing my strength would not last long much longer, this time he insisted I leave him and head off on my own. We argued vehemently but he would not be dissuaded. Before leaving I scouted around once again, foraging for food for the both of us. Ascending a nearby hill I saw a small shepherd's croft where a man was sitting watching his goats.

Approaching him cautiously I asked if he could help us. He agreed and together we managed to carry John into the croft. The man shared his food with us before helping John into the small bed. John could speak a little Greek and so we spoke a broken language describing to the man how we had missed being taken on the last ship during the evacuation and how we were on the run from the Germans.

The old man spat when he spoke of the Germans, cursing them with all his might. They had shot and killed his son and daughter. He was filled with hate for them. Knowing I could no longer help John, the old man said he would care for him, reiterating John's words that I should continue unencumbered on my journey north-east. He told me which path to follow and how far I needed to travel before I could expect to find the next little croft.

I continued to travel mostly during the night, sleeping in daylight hours. Suddenly awakening from a bad dream, I found the air was cool and there was the smell of rain. Raising the binoculars, I assessed the surrounding area looking for signs of enemy troops. Dead ahead there was a group of six Huns standing in a clearing about fifty yards away. After a prolonged discussion they headed right for my position. Soundlessly I took a couple of quick defensive steps backwards before melting into the trees and the falling darkness. The nightmare had saved me.

The following morning I came to a small clearing and looked around. On one side was a small farmhouse surrounded by a low stone wall. Just along from the gateway were small outcrops of rocks and stones piled one on top of the other. Within this formation was a small niche containing a small religious statue. I knew by the way these rocks were arranged it was a signal letting Allied soldiers know this was a safe house. Behind the statue I found some bread and figs. As I hungrily ate the food an old woman came from the back of the house gesturing for me to quickly come inside.

The atmosphere inside the small house was warm and welcoming. The woman and her husband encouraged me to stay with them. When they said they would hide me in the cellar I was doubtful. The cellar door was clearly visible over beside the kitchen sink making it a prime spot for the Germans to search if they came to the house. The old man smiled, shook his head and laughed. He beckoned me to the corner of the kitchen where underneath an old rug partially hidden by a box of firewood was the entrance to another cellar. After discovering there were caves not too far away that I could escape to if the need arose, I accepted their generous offer to stay with the family. If I was to survive, I needed to stop in one place long enough to regain my health and strength.

They kept pointing to my skin and then at their skin. My Aboriginal skin and hair colouring were dark though not black, so blending in with the local population would be easier for me than it would have been for any of my white Australian mates who were also on the run. My skin colouring worked in my favour making me appear of Mediterranean descent.

Contact had been established early in the piece with members of the resistance. Many nights I would join them in forays against the Germans. These attacks were often carried out well away from where I was hiding. If the family was caught harbouring me they knew no mercy would be shown and they would be shot. It made no difference, they protected me anyway. During the day I would disguise myself as an old man working in the field or attending to the goats. It was many months before arrangements could be made with the resistance to move me closer to the coast. Once there I would be smuggled aboard a small fishing boat and taken to Egypt.

It was very hard saying farewell to these kind, generous people who risked everything to keep me safe and well. They looked after me as if I was one of their own. I could not have asked for anything more. It was a great honour to enjoy the friendship of such brave and noble people. I would never forget them.

A single member of the resistance travelled with me as my guide to the coast. Once we arrived at our assigned point he shook my hand and wished me luck. With a quick nod of his head he motioned for me to continue, indicating the track, overgrown with weeds, which cut through the bushes leading to the beach. I lay hidden on the sand until the boat arrived then quickly swam out to it.

With the assistance of sympathetic and brave Cretans, I had avoided capture for eleven months. On 7 May 1942 I escaped aboard a trawler to Bardia, Libya. After surviving for almost twelve months I was on my way back to my unit and real freedom.

Tim and the final outcome

My mentor and good friend Angelo and I continued to work with the resistance during the German's occupation of our island. There are many stories of the brave members of the resistance who willingly put their lives on the line in order to help their fellow Cretans and the Allied soldiers stranded on the island.

In June 1942 the resistance were responsible for destroying 25 German aeroplanes. The German's answer to these attacks was to take 50 male hostages at Heraklion and execute them in front of the villagers. These unspeakable and indiscriminate killings of innocent civilians did not cause the Cretans to fold. Instead it fuelled their anger and determination to continue to cause as much trouble and confusion as possible for the hated invaders. It was their way of getting revenge for their people. They would never give in. There are humbling stories of families who time and again risked everything to aid the Allied soldiers.

At one stage there were over 50,000 Nazi soldiers on the island but they were not conquerors. Major General Ringel was so enraged by the actions of the Cretans that he ordered his troop of elite mountaineers to carry out reprisals against civilians who fought and killed the invading Germans. In one such attack he ordered the village of Kandanos to be destroyed. In retribution for the death of 15 Nazi soldiers, 180 citizens, men, women and children were shot and all buildings and livestock destroyed.

My good friend Angelo Bournazos was killed by the Germans in 1945. He was just one of thousands of brave, heroic Cretans who lost their lives. All over the island, you will find graves clustered around village churches or in tiny garden chapels. Their lives were given in defence of everything they held dear; they are the true heroes of this event in our history.

The Germans may have won the battle of Crete but the actions of the heroic Cretans contributed to them losing the war. Right to the very end the Cretan population continued to defy the Germans despite suffering the loss of over 20,000 of their own. These honourable people, who orchestrated the escape of many Allied soldiers, endured four long hard years of occupation.

The spirited resistance of the Cretans further compounded the delays experienced by the German Army as a consequence of supporting Italy in its bid to capture Greece. Hitler was forced to leave a large contingent of soldiers on the island in order to hold this hard fought and costly strategic position in the Mediterranean Sea. The loss of time and the massive redeployment of soldiers to Crete delayed and weakened his planned attack on Russia.

Without the strong resistance from the Cretans the Allies may never have won the Second World War.

CHAPTER 10

Life as POWs

A bleak future ahead

Every Allied serviceman knew the rules associated with being captured and held as a prisoner of war (POW) according to the Geneva Convention. The only things that needed to be voluntarily given to the captors were your rank and name.

Germans speaking perfect English carried out the interrogations and they tried many different ways to trip you up. Sometimes they pretended they knew much more than they did hoping someone would unwittingly divulge information, occasionally details were accidently leaked, though not too often.

Most of the chaps were very clear on what to do in these situations. After one or two goes at trying to elicit information from us the Germans would send us off to a transit camp or proper camp depending where you were captured. We would be farewelled with the words, "For you, the war is over!" This was the truth; our fighting days were over.

Ordinary servicemen were required to do any work they were instructed to, providing it was not dangerous and did not support or enhance the German war effort. For the most part the German army adhered to these rules. Senior NCOs (sergeants and above) were required to work in a supervisory role, while officers were not required to work.

Likewise German POW camps were separated into two types. An 'Oflag' was a prisoner of war camp where officers were held, whereas a 'Stalag' was for enlisted personnel. Within these camps the three branches of the military were usually separated into army, navy and air force and held in different sections.

As the war continued after 1941, Hitler was forced to recognise his war effort was draining Germany of manpower for its industry and agriculture. He needed every able bodied man (i.e. POWs) to work in these important fields.

Work performed by POWs was primarily undertaken in the mining industry, stone quarrying, sawmills, factories, railway yards, forests and farming. Local councils, military and civilian contractors all vied for workers from the camps. POWs were paid a very small amount of money for each day's work and generally received at least one day a week rest. These payments came in the form of 'camp dollars'.

Andrew Borrie - Crete

It was a struggle but I was successful in getting my ambulance load of wounded men aboard the evacuation ship in Greece. In the dark we headed out to sea, a group of us who were together in a huddle were from New Zealand. One of the men, a Maori with a smooth voice, broke out in song singing, *Now is the Hour*. It brought a lump to my throat – would we make it safely to Crete or might we be saying goodbye forever? Thankfully, we arrived safely and I could disembark the men and get them to the hospital and much needed treatment. I then found the No. 7 General Hospital in Crete where I had been attached for duty.

Over the next three weeks I spent hours each day transporting sick and wounded men from the incoming ships back to the hospital. When I wasn't driving I worked in the wards helping set up new beds or mattresses on the floor, taking men for assisted walks and doing general duties asked of me.

Just on daybreak before any reconnaissance planes were seen in the sky, I could be found waiting on the docks ready to retrieve medical supplies the salvagers had successfully brought to shore. Everything would be carefully loaded into the ambulance and I would be off, back to the hospital. The critical shortage of medical supplies forced us to improvise in order to source materials required by the doctors, nurses, orderlies and masseuses.

The local people were very good to us, and it was such a relief to get time off and mix with these hard working people. Those who came and helped at the hospital worked doggedly at their tasks, nothing was too much trouble for them. We had come to help them and they returned the favour in kind. Wonderful people!

Once the attacks started in earnest on the island on 20 May 1941, we were almost overrun at the hospital. The Germans didn't bomb the hospital, which was marked with the familiar red cross on the roof, but the fighting around Suda Bay was intense. Every day I was required to leave the safety of the hospital to collect the wounded from the surrounding areas. My heart was in my mouth every time my team drove out to help with the wounded. Transporting those in dire need back to the hospital as quickly as possible was demanding and draining.

In Crete I was not as fortunate as I had been in Greece. We could see our combat soldiers were losing the battle. With German forces landing more troops every day it became critical for us to save our nurses and any doctors who could be spared. While they all wanted to remain to care for our wounded who could not be moved, it was decided to call for volunteers to remain, as with most combat zones we had many men who were unable to be moved safely.

At the start of the war I had lied about my age, putting it down by five years. I had been too young to enlist in the First World War and was desperate to enlist in the Second World War. Now I was 41 years old with a few health problems and I faced the real prospect of becoming a prisoner of war. It was clear to me I would have difficulty crossing the White Mountains, so I opted to stay with those unable to be moved. We knew there would be no escape for those of us left behind but at least we could make sure our wounded were cared for. When the Germans eventually took control of Suda Bay, I was captured along with everyone who had remained at the hospital.

Captured

My poor family back home in New Zealand received word to say I was missing in action in July 1941. This was further confirmed by a report of Missing in Action published in the Otago Daily Times on 1 July 1941. It was not until October 1941 that they received official word declaring I had been taken as a prisoner of war on Crete and was being held in a transit holding camp.

Andrew and his wife

Our local newspaper, the Auckland Star printed an article on 21 October 1941 describing how relatives in Auckland had received confirmation through the International Red Cross regarding two prisoners of war. Corporal R.A.J. Bushell of Mount Albert and myself had now been officially taken off the Missing in Action list and were confirmed POWs in Crete.

On the move

After months of being held in the transit camp on Crete we were shipped to POW camps in Greece. On 24 November I found myself being herded onto a filthy cattle truck for the long journey from the bottom of Greece to Lamsdorf in Poland. A horror trip is the only way to describe what we endured along the way.

The roads were almost non-existent as a consequence of earlier bombing; food was scarce and warm clothing and blankets were not issued to us. Some of the men had no boots and others may as well not have for all the good they were. By the end of the journey many of these men needed treatment for frostbite. We lost a number of good men through illness and disease on this seemingly unending nightmare of a journey. Once a day the trucks would stop and we were allowed to alight under heavy guard.

We craved the meagre food rations we were issued and the space to move around a little. Death is not a pretty sight at the best of times, but it was particularly hard for us, especially being forced to stand in a crammed crate for hours on end with dead mates at our feet. During the daily break on our journey we were made to dig graves, unload the bodies of those who had died in transit and bury them.

Stalag VIIIB

Initially I was sent to Stalag VIIIB. It was a very large camp with between 8,000 and 15,000 prisoners at any one time. My POW number was 21824.

The camp was divided into several compounds with five barracks each housing 400 men in two divided sections. Each barracks had a large stove fuelled by wood or brown coal briquettes and this was used for cooking and heating.

Thankfully e ach c ompound a lso h ad i ts o wn m edical r oom w ith a British medical officer in charge. Nearby there was a prisoner-of-war hospital with a capacity of 450 beds. Upon arrival we were all sent to the delousing hut and from there to our barracks.

As I was a member of the New Zealand Medical Corp I was fortunate enough to be assigned work as an orderly in the hospital. The camp hospitals weren't much by way of our hospitals at home, but they had their fair share of sick and dying. At least our sick and wounded who were admitted to the hospital were given clean pyjamas and a blanket when they arrived. Water and drainage were a major problem at the camp, especially before they built a new bathhouse in 1942. Can you imagine 15,000 men all trying to get showered? On average I got to shower once every ten to twelve days. The main source of water was inadequate for the number of men held there and at times the water would be cut off for hours at a time.

Most of us only had the clothes we were wearing when we were captured and obviously these were not going to last long. Every POW work camp elected one fellow prisoner who was known as our man of confidence, to act as liaison between the prisoners and camp authorities. Often this would be someone of a senior rank in the camp, but not always. Our man of confidence was constantly harping at the Germans to supply us with clothing particularly as some of the men's uniforms were in terrible states of disrepair. Eventually some ration controlled clothing was issued to us. I believe some of this was obtained through the International Red Cross. We were allocated British army boots, shirts and woollen underclothing.

Over the coming months we received irregular deliveries of Red Cross packages and eventually some parcels from home. Everyone looked forward to receiving these, because apart from news of our families they would often contain food and additional items, including things like woollen gloves, scarves, hats, underclothing and books. We never knew when we would receive another allocation so we tried our best to look after our clothes and make them last.

On occasions some greatcoats and gloves would be supplied, for obvious reasons these would be first given to the men who were assigned to the working parties. Any surplus was then divided up between those who required them. Some of the poor devils sent out on the working parties in the freezing cold would arrive back at the sick bay in a terrible state. I always found it upsetting caring for those suffering with frostbite of the fingers, toes, nose, ears, cheeks and chin.

By the time the medico was able to inspect the joint or exposed flesh the skin may have already changed colour to a bluish grey and numbness or loss of sensation to the area would already have set in. Often the men had tried to warm their hands in hot water or over the fire, which would lead to large blisters forming within hours. In extreme cases we would see areas where the skin had turned hard and black and the tissue had died. The doctor's only option then was to operate and remove fingers, toes, parts of the ears and sometimes the tips of noses. It was shocking, just shocking!

It was difficult to see the men who were suffering from the 'horrors'. You could see the effects of flashbacks play across their faces hurtling them back into situations they would never be able to reconcile. They would jerk forward, jump up and run headlong into their nightmares. The loss of a mind, something destroyed by war, the pain experienced behind those empty stares where no one could reach.

The Germans provided a daily ration of barley porridge made with hot water in the mornings or a watery potato and cabbage soup in the evenings. Heavy black bread and a piece of rough German cheese were usually issued one between two men. Thank goodness for the Red Cross food parcels and parcels from home; I don't know how any of us would have survived internment without them.

In 1942 I spent time at Oflag IX A and in 1943 I was transferred to Oflag IX A.4. This was to be the place of my last internment. My age and health were against me and I ended up spending time in hospital as a patient myself. Because of my failing health I was chosen for a prisoner exchange and repatriated to Egypt in November 1943. From Cairo I was sent back to New Zealand on 31 December 1943.

I never forgot the bone chilling cold in winter. I could never seem to get warm and developed persistent bronchitis and lung problems. All in all I guess I didn't fare too badly under German imprisonment and at least they gave us blankets and soap. I will never forget the near starvation and cold endured during the nightmare journey to the camps nor the deprivations of captivity.

Douglas Channell – Oflag VIIB

After being captured on Crete on 31 May 1941, I was eventually sent to Oflag VII B Colditz, Bulgaria, where all prisoners were kept in chains for eighteen months.

My fellow inmates, Kenneth Haig MacQueen, Patrick (Pat) Laurie and Douglas Bader became good friends. While not required to go out on the working gangs, Pat and I were granted permission to go and collect firewood for the barrack stoves. As officers it was our 'duty' to try to escape if possible. One of these methods involved tunnelling under the camp.

The gathering of firewood gave us an opportunity to size up the lay of the land and look for possible spots where an escape tunnel could end. Even though the possibility of escaping from Colditz was very slim it still gave us something to strive for.

While escapes were always on our minds, we also had the men maintain their routine of daily procedures. The Germans would conduct role calls, checking to make sure no one had escaped through the night. The men would complete their daily drills and maintain their sense of order. We were still highly trained soldiers with rules and routines to follow even if we were POWs. Routine gave us something to hang onto, something to take pride in and something that pissed off the Germans no end. We would not give in; they could not break us. As a group we were still a formidable foe.

On a lighter note we formed a very workable entertainment program for ourselves. The rules of the Geneva Convention maintained all POWs should have certain rights. To this end we performed and participated in plays, concerts and sports. One of the more educational items on our agenda was to give lectures to the men in the other sections. We had plenty of time on our hands and many of the men were skilled craftsmen in their own right. We would organise classes where those wanting to learn more about a particular form of work would join a session given by someone who had experience in that area.

As a radio announcer, I could coach the men playing parts in a play how to learn their lines, how to pitch their voices, use different tones, alternate between playing the voice role of a man and then swapping to that of a woman. Everyone happily shared their skills and stories of their families with each other. After all we were on our own, and as I said we had plenty of time on our hands.

Everyone participated in our escape plans. Digging tunnels was long difficult work. We put systems in place on how to get the dirt from the tunnels out without being discovered. We found if we cut a hole in our pockets and tied them up with string then we could fill our pockets with dirt. This was done just before we went out to play sport. Once on the field we would release the string and allow the dirt to fall out on the ground. With all the running around the guards never realised what was happening right under their noses.

Timber was required to shore up the tunnels and we found removing a single slat from each of the beds gave us an initial supply of wood. Sometimes when foraging for firewood we would pick up planks from the surrounding areas and appear to cut them up for fire wood. We might do this with one or two but the rest went down the tunnel or were hidden from the guards until needed.

I stood six foot three inches and was a big man to boot so there was no way I could ever get through the small tunnels. Added to this, I still found it difficult to use my left arm properly due to the shrapnel wound in my back. Besides it would stick out like a sore thumb if I was not in attendance at roll call as I was the tallest man in the camp. Had I contemplated escaping, it would have been downright impossible to keep my disappearance a secret.

Camp activity was geared for escapes. While some of the men concentrated on building the escape tunnels the remainder of us worked on other aspects of the escape plans.

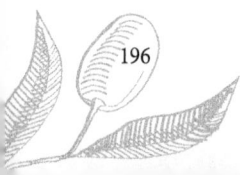

Everyone had their own particular skills to offer, the tailor would organise some of the costumes to resemble civilian clothing, someone else would work on forging documents and secure maps, someone organised photographs for fake documents, while others who had learned German were used as ferrets to distract guards if they became curious about any activities related to our escape plans. Everyone worked together with a united plan.

The Germans did not have any form of entertainment, so whenever we held our plays and concerts they were always keen to be invited. They certainly did not realise that as well as providing entertainment for everyone we were also quietly working on our escape plans. We also enjoyed it because we could openly take the mickey out of them without them understanding our humour.

Double amputee Doug Bader never let the fact he lost his legs in an air accident stop him from planning escapes, he was a good man and he thought of himself as invincible. He wanted to escape so badly, but unfortunately on one of the planned escapes he was too slow crawling through the tunnel and inevitably held everyone else up. This delayed them to the point where they were forced to abandon the night's planned breakout.

One morning, early in 1945 while walking around the compound I decided I had endured enough. Two German officers passed me and I refused to salute them. When they pulled me up I still refused to salute, telling them I didn't care to salute a lot of bloody murdering bastards. Well that little episode ended in me being court martialled by the Germans. During my trial I was accused of insulting the German army. My reply to them was,

"What the whole lot of you, aren't I clever?"

There were a number of us on trial that day and when the man beside me laughed and chimed in with another insult they shot him dead for his insolence. We were all so shocked; we couldn't believe our eyes. I was right, they were a bunch of murdering bastards. They made their judgement and I was sentenced to solitary confinement for two years.

Solitary confinement consisted of being housed in a very small, confined cell, no easy task for a tall man like myself. I couldn't stand up straight and had no room to move around. The only way I could get any relief was to lie on my back on the bed resting my legs on the opposite wall and stare at the ceiling.

In the first few weeks, I thought long and hard about the murder of the man standing beside me during the trial. A strange feeling had been growing in me and I couldn't get it out of my head, it didn't make sense. It just went around and around—why did God let this happen?

Englishman Padre Tony did an outstanding job of keeping the men's spiritual sanity in place; not an easy task in this hellhole. But at the time, try as he might, there was no way the Padre could help me deal with the murder of the soldier at my trial, I had lost my belief in God.

The boredom, endless dark hours, inability to speak with anyone, loneliness and loss of faith did take their toll on me, though there was no way I would let the Huns know that. My way of dealing with things was to always use humour and this is how I coped.

I was allowed out of my cell to exercise for one hour twice a day. An armed guard was assigned to watch me the whole time. I took full advantage of this precious amount of free time to move, stretch, exercise and breathe in the fresh air.

If they were going to keep me locked up, then I would make them work for their time with me. Being six foot three meant I had a long stride and I would walk very briskly making some of the guards almost have to run to keep up with me.

It pissed them off when I would come out into the exercise yard cracking jokes and singing as if nothing had happened and everything was normal. If they were going to punish me then I was going to give them something to think about as well. It was too darn bad if they didn't like what appeared to them to be my pompous attitude. I got a lot of enjoyment out of their anguish and this only encouraged me to sing louder. It drove them crazy realising I had no intention of cowering down to them.

Although we were prisoners of war and some of the guards treated us badly, there were plenty of Germans who were kind and treated us well. This included some of the locals.

One morning in April 1945 the Germans suddenly disappeared from the compound. My friends released me from my cell and told me the good news; the war and our confinement were over. When the Germans had first allowed us to walk through the town under guard, a local man called Peter Biebel, was very good to me. Now we were free, we once again walked through the town. I visited the home of Peter and while I was there I wrote a note saying, "Please look after this man because he looked after us." Men with me witnessed this note and signed their names to it. It would keep Peter and his family safe once the Allies arrived.

While we waited for our rescuers to arrive some of our planes flew over. They started to strafe us thinking we were Germans. We ran and hid in the furrows of the field. Thankfully that was not repeated as our rescuers arrived soon after.

After release, I was preparing to be repatriated to a transit camp in England. Six years in the army, four of those as a POW meant I was well and truly ready to go home and reconnect with my family. While waiting at the airstrip to board the flight back to England I witnessed yet another horror, one of the planes loaded with soldiers crashed. Unbelievably, after waiting so long for freedom many of the men were killed on impact. I longed for the insanity of war to end.

Upon arrival in London, I phoned my parent's home number, when the woman on the switchboard answered, I explained I had no money but I was a returned soldier trying to contact my family. She told me, "You don't need money, I'll put you through." I had not seen or spoken to my family since leaving home at the age of fourteen, so when Dad answered the call I told him I was his son. Dad asked, "Which one are you?" "I'm Douglas." We organised to meet at the train station. When Dad laid eyes on me he said, "You are a big bugger, aren't you!" He took me in hand and we went home where Mum was so happy to see me after so many years.

I stayed home and worked for the BBC until being called back to Australia to take up the reins at the ABC. Once again I was 'The Man on the Street', and I also delivered a program called 'The Housewives Choice'. I married and had twins, Peter and Lesley who were with me on the day I was presented with the Military Cross. Career wise I went on to become the Queensland manager for the ABC.

John Borrie – Doctor, doctor

I was captured in Corinth, Greece on 24 April 1941. The Germans came looking for doctors. The Red Cross badge stitched on my sleeve and my surgical haversack under my arm left them in no doubt that I was a doctor. After rounding up myself and Captains Foreman, Neale and Slater, they transported us by ambulance into Corinth. I thanked God I was alive and thought how lucky I was to be a doctor. The Germans wanted us to start up a British prisoner of war (POW) hospital.

Ariadne Massautti & the Greek Red Cross

The Greek Red Cross had secured the Ionian Palace Hotel as the location for the hospital. At first there was some confusion as to how we could achieve setting up a hospital with no drugs, equipment, dressing or beds. That was until I met the 'little old lady'. I will never forget the first time I laid eyes on Miss Ariadne Massautti. This little old woman had such a powerful persona the moment you met her you realised anything was possible. This seventy-odd year old woman had persuaded the German paratroop's Dr Bauer to allow her to use the Ionian Palace Hotel as a British POW hospital in the name of the Greek Red Cross Society. When some of the Germans tried to take over the facility for their own wounded she fought back and had all the wounded Allied soldiers taken from other Greek hospitals and transported to the hotel POW hospital.

In no time we had 50 men arrive. I grabbed the hotel register and used this as the admissions and discharge register. After an inspection I knew there were 120 beds but no fresh linen, medical supplies or food. A steady stream of men from the 4th Hussars, Australians and New Zealanders began to pour into the hospital. The situation looked hopeless. While we were trying to get our heads around what to do, we heard a commotion out the front of the building. Suddenly a large group of Greek women entered carrying baskets of food, piles of fresh bed linen, dressings, antiseptics, instruments and primus stoves.

This was indeed a sight for sore eyes. The genuine willingness by these women to help us to begin to treat our patients with some degree of success. Dr Bauer left us after things started to fall into place. I will never forget him as he was the best German I had ever met.

The worst cases were placed on the beds while the less urgent cases were put on mattresses on the floor. More medical staff, including a small number of dentists and around ninety orderlies arrived in dribs and drabs. Every time the Greek Red Cross arrived our small supply of dressings increased noticeably. Ariadne was instrumental in securing space at an operating theatre at one of the Greek hospitals for our use.

On our way to the hospital, Greeks would give us the thumbs up sign. Even though they had nothing to cheer about they were still happy to show their support for us Allied POWs.

Illness & injuries

Over the next four years I found myself working at a number of different camps. We were a mismatch of an army, weakened by many different aliments, including over-work, starvation and infections. Many of the men were just walking skeletons. Men would toss and turn, bathing themselves in sweat, fighting disease or failing to stop reliving events they tried so valiantly to put out of their minds during daylight hours. Seeing the despair, frustration and exhaustion in the men would give my medical staff the determination to do whatever we could to help these broken souls. Mercifully as time in the camps turned into years many of the men learned mechanisms to help heal, or at least deal with, their wounded minds.

Some of the working parties were tough. Men would return to camp from the mines after spending the day clawing their way up hillsides, dragging loads of rocks through mud, snow or driving rain. Their fingernails were ripped from their fingers, bodies exhausted from the effort, needing, craving rest but there was no rest. The guards would make them unload their burdens, turn around and do it all again.

Many of the Germans were fair and allowed us to live a more civilised life, well as much as one could while spending years cooped up in Stalags. Whenever things started to get me down I would bring to mind the amazing work carried out by Ariadne Massautti and the very kind hearted women of the Greek Red Cross. These thoughts would give me the strength and courage to continue.

I was held captive for four years. Towards the end of our confinement we had the Americans to the east and the Russians to the west of us. The Germans were in a pickle not knowing which way to turn. They organised to transport us by train but the train failed to arrive, so they decided to march us all to Nuremberg. We were caught in the middle with bombing and strafing coming from our Allies.

It was Easter on 30 March 1945 and thankfully while all of this was going on Padre King continued his Holy Week Service in his little chapel. Captain Foreman argued with the camp commandant that our men were incapable of being moved unless they were taken in proper ambulance trains. He refused point blank to move them. It was settled we were to stay where we were.

On 5 April we waited in the miserably cold weather, apprehensive for the safety of the men still in our care. Looking out the next afternoon we could see soldiers dressed in khaki uniforms holding sten guns. Our rescuers had arrived. It was Friday 6 April 1945.

My diary entry for that day read:

> I find it difficult to write today, strangely quiet, yet exhausted, as though on the edge of a new and wonderful world. At 3pm my captivity of four years less 20 days ended – Gott sei Dank!

Hitler used his secret weapon of 'neglect' when dealing with POWs and I can honestly say if it had not been for the International Red Cross Society, with its life supporting food and clothing, we would not have survived our four years as captives.

Clifford Morris

After being stranded on Crete I was able to keep hidden until I was able to make contact with the resistance. For two and a half weeks I worked with them to hinder the Germans and to find and locate others like myself who were evading capture. One night we were heading out to meet up with another group at a barn two miles from where I had been staying. I don't know what happened as I was towards the back of the group when suddenly there was shooting. We had walked right into a trap. Two of the men with me were killed. I was fortunate to be unharmed but devastated when I realised my freedom had come to an end.

I was placed in a holding camp for a few weeks and then shipped to Greece. Then began the long trek to Austria. This was accomplished by various means of transport—on foot, by truck and train. The railways and roads were in a dreadful state after the damage from the bombings. I eventually ended up being incarcerated at Stalag XVIIIA. Our few personal possessions were precious to us as they were our only link to home. Home was so far away and we clung to any item that gave us hope to keep going. A treasured book or a photograph; letters from home; a small carving lovingly made from a scrap of timber; scavenged, irreplaceable items helped us remember we belonged to and were loved by someone. Thoughts of home, though at times painful, were a powerhouse of incentive to keep going. We were going to get through this and we were going home someday.

It was a real shock to go from being free to being locked up for who knows how long. We were thousands of miles from home, food was restricted and there was nowhere to go except in circles around the exercise ground. To give up would have been so easy but we were better than that. Sadly a few of the men couldn't deal with with being POWs but for the most part we were survivors.

Life quickly fell into a routine. We would try to wash in cold water once a day and shave at least every other day. Our laundry was carried out weekly if possible. Sometimes we would receive soap in our Red Cross packages otherwise we would have to resort to using the German 'ersatz' soap.

Any of our men who died whilst in captivity were honourably buried in individually marked graves. One morning after having just come off burial detail, I took the opportunity to volunteer to work on one of the farms not too far away.

Working life as a POW

Many of the POWs at each stalag were assigned to smaller work camps or local sub-camps. For the most part this required us to work and live at the smaller camp so we could undertake our assigned duties at factories, coal mines, quarries, railway maintenance or local council worksites. Sometimes men would be assigned to work on a farm and would be housed at the farm with the farmer's family.

This meant I would be assigned to work and live at a smaller camp attached to the farm. I wanted to escape if possible, and knew it would be impossible to escape from the main camp at Stalag XVIIIA especially with twice-daily parades. I thought my chances of escape might be better from a smaller camp, so I stuck up my hand when they asked for volunteers.

Escape

Before leaving the main camp a mate and I had spoken to the NCO letting him know what we had in mind. He organised two bundles of civilian clothing for us from the entertainment costumes.

A small group of us left camp with a great deal of hope and expectation. After marching for eight hours we eventually arrived at a farm high up on the mountain in the forest. Our job was to cut timber. The guards had travelled with us and also stayed at the farm. The farmer and his wife were very kind to us and we were well looked after. Even though we were housed in the barn we had plenty of warm bedding and good food. I became friendly with one of the guards. He was a good man and now we were away from the main camp our friendship grew. He didn't want to be there any more than I did.

The work was hard but it was so much better than being stuck in the main camp with thousands of other soldiers. After months of living in the forest we were healthy, well fed and fit. One day while we were cutting down a massive tree, one of the men lost his footing just as the tree started to fall. He was badly injured and had to be transported back to the Lazarett (Hospital) at Stalag XVIIIA for treatment. This required two of the guards to go back with him. This unexpected event provided us with an opportunity to effect an escape. I had become friendly with a guard named Koloman who organised maps and false papers for us on his visits to the nearby town on his days off.

At the end of the day's work, Koloman and the remaining guard went to their quarters after dinner, leaving us alone for a period in the shed. Quickly my mate and I changed into the civilian clothing, grabbed some food we had hidden and took off.

We knew we were close to the Swiss border and set a course to the quickest way to reach a point where we could cross safely. Our main problem was it was the start of winter and there had been heavy falls of snow. We had no experience of moving in the snow and it didn't take us too long to work out the hopelessness of our situation.

If we ran into a group of Nazi Alpine guards on skis we wouldn't stand a chance. Most of those Nazis were of Olympic standard and we would be sitting ducks. Hard though the decision was, we opted to return to the farm and wait for another opportunity.

We arrived back at the farm just before dawn, exhausted but ready to put in another full day's work before we could rest our weary bodies. Two weeks after this attempt we were moved back to the main camp as the snow was too heavy for us to be able to work. Our opportunity to escape was now dashed for good.

Another escape attempt

Back in the main camp, I volunteered to work on the escape committee. We were lucky to have a carpenter in our hub who worked on securing timber for the tunnels. He took slats from the bunk beds, leaving just two under our head and two under our bum and one for our feet. Red Cross parcels arrived in wooden boxes and while he made some into seats the balance of the timber was used in the tunnels. Other men were responsible for making sections of long piping to provide the men digging the tunnels with circulating air.

My next attempt at escape ended in disaster. I was recaptured and another Aussie mate, Alan Eason, said to me afterwards, "Geez mate I don't know how you weren't lined up against the wall and shot." That is something I myself will never know. After facing a court martial I was put into solitary for two months but it didn't stop me dreaming of escaping. I never told the Germans where I had got my maps or false papers. Koloman was a good man there was no way in the world I would ever betray his trust.

CHAPTER 11

Death from friendly fire

Alan Eason - Footy Hero

I was born in Brisbane on 12 November 1917. Dad was offered a position in Warwick and the family moved there. As we lived close to the middle of town I completed my schooling at Warwick Central State School. Upon my graduation from school I secured employment with the English, Scottish Commercial Association in Brisbane as a Commercial Traveller. Living in Brisbane with my older brothers and working for the firm allowed me to indulge in a great social life and enabled me to follow a life time passion for footy. The youngest of four boys, it was natural for me to grow up with a love of rugby union. My older brothers and I all played rugby for Eagle Junction Aussie Rules Club.

Whenever you picked up a newspaper in those days, the sports section would often have reports on how one or other of us Eason boys had played a smashing game of footy over the weekend. All of us loved the game and became talented Aussie Rules players, representing not only our own club but also having the opportunity of trying out for the state side. Not long before I enlisted I was selected to represent Queensland in the Rugby Union Team of 1939 as a centre. Queensland defeated New South Wales 97 to 84. What a thrilling result for us as a team and for a young man like myself.

QLD Rugby Union Touring Team
Alan standing in the back row, third from the left
Photo courtesy of Eason Family

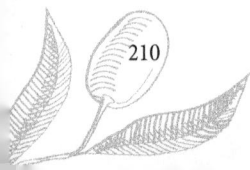

In 1939, with the outbreak of the Second World War in Europe, Australia's Prime Minister, Robert Menzies, committed Australia's 6th Division Second AIF to the Middle East under the command of Lieutenant-General Sir Thomas Blamey. My unit, the 1st Corps Troops Supply Column, Eastern Command, 2nd AIF made up part of the 6th Division.

It must have been hard for my parents to see all four of their sons join up for the Second World War. My older brothers were Raymond Cyril Eason QX39435, Keith Alexander Eason QX6780 and Colin Leonard Eason Q126659. I wanted to join my older brothers and do my bit for the war effort. So on 29 July 1940, I enlisted and was recorded as Alan Ernest Eason QX17451. Poor Mother, it could not have been easy for her to wave all four of us off on our way to fight.

I still recall my father, Cyril Eason bidding me farewell at the front gate of our home in Warwick and wishing me well. At the time I had been home on a seven-day pre-embarkation leave after completing two months of intensive training in Brisbane. I needed to leave home that morning as I was due to march out to Sydney with my unit on 2 February 1941.

On the move

Arriving in Alexandria in February 1941, 26 years after those other young ANZACs had during the First World War, was a surreal experience to say the least. During our time here the unit undertook more training before we were once again on the move.

We left Alexandria for Greece on 4 April 1941 to join other units under the command of the 6th Division to assist the Greek Government in their endeavour to fight off the German and Italian onslaught directed at their country. This campaign was to be a short one for us as the advancing German army had a far superior number of troops, equipment and, what seemed to us on the ground, unending air support. The Allied forces were quickly overcome despite spirited fighting by our troops. A masterful plan of withdrawal from Greece was quickly and efficiently executed under difficult conditions and ended up saving the lives of thousands of Allied troops. What was left of I Corps HQ left Greece between 23 and 24 April and were returned to Alexandria less than three weeks after our departure from there.

Lady Luck goes AWOL

Regrettably I was not amongst those who returned to Alexandria. Lady Luck was not on my side. I found myself fighting and for a short time I remained at large, but eventually like many Australian, New Zealand and British soldiers I was captured and taken as a prisoner of war. Still I was relieved to know a small number of men who had been with me, managed to escape to the hills and continued the fight, assisting the Greek partisans in the struggle against the marauding Germans.

On 10 May 1941 it was officially reported that I was missing in action in Greece, believed to be a prisoner of war. My name was then transferred to the X List. Meanwhile, at home in Warwick, Dad received the dreaded and unwanted telegram from the army. That little piece of paper, he later wrote to me, was enough to stop him breathing.

A seemingly insignificant envelope could make or break a loved one. Did they dare to open it? If only it hadn't arrived, one could continue to live in the eternal hope things would be all right. What would he tell Mother? The telegram that arrived for Dad was both good and bad.

POW in Stalag XVIIIA

Initial reports stated I was missing in action, believed killed, but for my parents there was always a glimmer of hope I might be found alive. The telegram informed them the International Red Cross confirmed I had been taken prisoner and transported to a German Prisoner of War Camp called Stalag XVIIIA in Austria, on 23 June 1941. The family thanked the Lord, their prayers had been answered and I was still alive.

After my initial capture I was placed in a holding camp with other Allied troops before being sent to Stalag XVIIIA at Wolfsburg, Austria. This was a truly gruelling journey of many weeks, involving forced marches and transportation by train. The train carriages had previously been used to transport cattle. Our captors hadn't even taken the time to wash the carriages out before shoving us inside and locking the door. The outcome meant we were forced to travel standing in animal excrement with only one small open gap at the top of the carriage to let light and air in. Locked in and confined within these filthy, stinking spaces was inhumane to say the least. Conditions were dreadful, we were overcrowded, near starving and had only one tin bucket in the corner to use for sanitation purposes. Dysentery among the men was rampant after a few days but with only one rest stop a day where we were let out of the carriages under heavy guard so we could relieve ourselves, it was not surprising a number of the men succumbed to the horrendous conditions. Demoralised already because of our capture, we were further traumatised when those who were too weak and sick simply lay down and died at our feet.

As we were unable to wash it was not long before we also had to contend with being covered in lice. All we could do was get the man behind you to pick off the lice on your back and you would pick the lice off the man in front of you and so it went. The entire trip, which took several weeks, was a nightmare and an experience I never wanted to be a part of again.

Things did not improve upon our initial arrival at the POW camp. Before we were given something to eat or allowed to clean ourselves up, we were forced to wait in line to be processed and issued with POW numbers. Our paper work then had to be forwarded via the International Red Cross to Allied Headquarters confirming capture and present whereabouts. Here I was, POW number 4035 at the ripe age of 23, starving, lice ridden, unshaven, filthy and incarcerated by the enemy in a foreign land 'til only God knew when.

This outcome was a far cry from what I had envisioned when saying farewell to my parents in Warwick, for now, I was a POW. From now on I was a POW of the Germans.

Work camp

Close to the main camp of Stalag XVIIIA there were two satellite work camps, these were 942GW and 10029GW. I opted to volunteer for the working party at 10029GW situated at Klagenfurt. It was a bigger camp, but much more organised and not as overcrowded. Work consisted of a squad going into the local town most days, where we would work as labourers for the local council. Camp vouchers to the value of ten cents were issued for each day's work and we could use these to buy goods and food.

There where over 900 Australians imprisoned here for the duration of the war and like the others I found it very depressing. However despite the many deprivations we endured, the prison guards here treated us reasonably well. At night there would be roll call and each of us would be marked off as being present, we were then ordered into our barracks and locked in until the morning, when once again we would line up for roll call and be marked off, the food was disgusting; watery boiled potato soup with no salt or meat, boiled cabbage and there was no such thing as tea, milk or sugar. How we managed not to starve is beyond me.

The general consensus was every POW had a duty to escape, but that was near on impossible here. We were right in the heart of German controlled land with nowhere to run to. Armed guards outside our locked huts, high barbed wire fences, sentries walking the boundaries with dogs and machine guns in the towers with lights put any thought of escape well and truly on the backburner. To top this off the Austrian people at the time were very much on the side of the Axis Powers and would have turned in any Allied soldier at the drop of a hat.

God bless the Red Cross

Back home in Warwick I had often complained of the cold in winter, trust me, it had nothing on the bitter winters at Klagenfurt. Snow, bitter winds, rain, limited warm clothing and very little heating in the huts left a feeling of being cold to the bone most of the time. The only good thing about winter was the work parties were often cancelled because of the bad conditions. The guards didn't want to stand around in the cold rain and snow watching us.

Sometimes in life you come across something that makes you grateful for the little you had. Close by there existed another prison camp that we could see through the barbed wire fences. It was filled with Jews and others considered below the lowest of the low by the Germans. These poor devils were treated in the most appalling way with no regard given to their person or health. For them there was no hope, they were considered expendable and worthless. It was like watching a group of skeletons walking around in rags. Compared to them we were very well off.

Arrival of, and access to, Red Cross parcels was a godsend and definitely contributed to our improved health and mental conditions. Our nominated 'man of confidence' would insist on being present when these parcels entered the camp. He and one of the German guards would store these packages in a secure place. The success of this system made sure we received the full contents from these parcels, which helped sustain us. How we looked forward to opening the packages and digging into the corned beef, spam, fish, packets of condensed milk, sugar, margarine, biscuits, tea, coffee and dried eggs. These little luxuries often meant the difference between near starvation and maintaining reasonably good health. When parcels from home arrived they would often include items of clothing and the all-important chocolates and sweets, something you were guaranteed could be used for bartering purposes if the opportunity presented.

Because we were taken as POWs while on active service we were also entitled to cigarettes. As most of us smoked we looked forward to receiving our allocation of 50 cigarettes (smokes) each per week. As a smoker I always took my full allocation and enjoyed each and every drag. Those who didn't smoke used theirs to bribe the guards or traded them with other prisoners or the locals for other goods and food. In all honesty the guard's food, though greater in quantity, was not much better than what we ate, so bartering with them was usually achievable despite being forbidden.

An example is how in exchange for one of our cigarettes the guard would secure a fresh egg for us. Something as simple as a fresh egg, an everyday item you wouldn't even think about doing without before the war became a most precious luxury while living in the POW camp.

In 1942 I wrote home with news of how I was coping and to acknowledge just how appreciative we all were for the important work carried out by the Red Cross. Without the Red Cross parcels we received, life in camp would have been all the harder to contend with.

RED CROSS TRIBUTE FROM WAR PRISONER

WARWICK. Thursday. – A tribute to the work of the Red Cross Society has been paid by Pts. Alan Eason, former interstate Rugby Union player and captain of the Eagle Junction team, who is a prisoner of war at Stalag 18 prison camp, Austria, having been captured by the Nazis in Greece in April, 1941. In letters to his parents – Mr. and Mrs. C. Eason, of Warwick – Pts. Eason writes that he cannot speak too highly of the work of the Red Cross Society, whose parcels are received regularly by the inmates of the camp. Though he works hard and the hours are long, he says that he is putting on weight and is in splendid health. A photograph of a group of 16 prisoners accompanying the letters show them to be apparently in good health.

Source: The Courier Mail: Fri 19 June 1942 / page 4

Entertainment

The middle of camp 10029GW was set aside for recreational purposes and it was here I was able to indulge my passion for footy. When allowed exercise breaks you would find me off with some of the lads playing a rough game of footy. There were also many organised games like athletics, tennis, footy, cricket and boxing.

The camp had a mixed population of Australian, American, New Zealand and Polish prisoners, and the senior service personnel among the prisoners organised entertainment such as drama plays, football and volleyball competitions, and a 'Commonwealth Games' in athletics in which I won a 'gold medal'.

Footy group - Back: John O'Malley, Lofty Windsor, Rob Adams. Middle: George Richardson, Ted Bradburn, Alan Eason. Front: Eric Green. 1943. Alan Eason and Eric Green were killed in the bomb raid on 19 Feb 1945. Photo courtesy Ian Brown www.stalag18a.org

In the four years I was incarcerated in Austria, I struck up many close friendships. Two of those friends were British POW, John Slack and fellow Australian soldier, Bill Cassidy. Together we suffered the deprivations confronting all prisoners at that time. We shared whatever we had and like all the prisoners looked forward to liberation.

While we worked long hard hours we also played hard during our recreation periods and had a good time mixing with the other prisoners.

John was very popular with the men and would keep us entertained in his role of helper of theatre productions. When the plays, pantomimes and skits were developed they often required some of us men to volunteer to play parts. Initially there was a fair bit of reluctance to participate, particularly if the part required someone to dress up as a woman but once things got going and everyone saw what fun could be had, it didn't take long before John and the entertainment crew had volunteers galore.

Countless men were sourced for their pre-war skills of building, painting or tailoring and they became very creative by making costumes, setting up stages, props and designing a myriad of painted scenes. Those who played musical instruments would get together to practise with real or made up instruments, often learning the songs from memory. Records and a record player were sourced with our 'camp money'.

CINDERELLA

A PANTOMIME
PRESENTED BY
"THE EIGHTEEN DEES"
AT

STALAG (XVIII D)   DEC. 1941.
CHARACTERS IN ORDER OF APPEARANCE

CINDERELLA	BILL HUTCHINGS
UGLY SISTERS { "CLARA"	LEN SUTTON
{ "AGGIE"	JOCK HUGHES
PAGE	BASIL SMITH
FAIRY GODMOTHER	FRED LANG
PRINCE CHARMING	RAY MASTIN

ARRANGED AND PRODUCED BY F. D. (NIC) NICHOLS
ASSISTED IN PREPARATION OF BOOK L. (DAD) SUTTON
SECRETARY AND TREASURER L. (UNCLE) COWAN
ORCHESTRA UNDER THE DIRECTION OF
DENNIS WHITELEY (PIANO)
LEW STAFFORD (VIOLIN)   ARTHUR RILEY (TRUMPET)
DICK DAVIS (GUITAR)   ROY CROPPER (DRUMS)

SCENE 1.		KITCHEN
"SYMPATHY"		CINDERELLA
SCENE 2.		SISTERS BEDROOM
"TWO UGLY SISTERS"		THE SISTERS
SCENE 3.		KITCHEN
"WISHING"		CINDERELLA

10 MINUTES INTERVAL

SCENE 4.		BALLROOM
MEMORIES LIVE LONGER THAN DREAMS		PRINCE
LOVE WALKED IN		CHARMING
SCENE 5.		STREET OUTSIDE KITCHEN
SCENE 6.		KITCHEN
STAY IN MY ARMS CINDERELLA		PRINCE
SWEET MYSTERY OF LIFE		CHARMING

STAGE MANAGER	JACK L. MASON
ASSISTANT STAGE MANA-	
GER	REG. COOK
LIGHTING	J. L. MASON, FRED DENHAM
COSTUMES	JOCK CUMMINGS
SCENERY	JIM WELCH
PROPS AND EFFECTS . . .	REG. COOK, TOM BAKER
PROMPTER	JIM KALEY
VOICE	GEORGE DUNNING

ACKNOWLEDGMENTS TO THE GERMAN CAMP STAFF
AND UNTEROFFIZIER W. LUDWIG FOR THEIR VALUABLE
COOPERATION

Example of a concert program

Productions were time consuming, particularly if you had a part and had to learn your lines, but oh it was so worth it in the end. We even had the guards sitting in the audience. Some of the shows would have us in stiches. Not only were they very enjoyable and risqué but often times double edged. We could make fun of the Germans without their knowledge, as they didn't understand our double-edged comments. Participating in these types of activities was a good way to pass the time and increase our knowledge and skills for when we were hopefully victorious and rescued from our internment.

My first passion though was sport. As a representative rugby union player I was always keen to have a kick around with the other lads. Every chance we got we would organise a couple of teams and have a game. It was a great way to let go of the stresses we endured day in and day out. I enjoyed coaching others on the finer points of the game and was proud of how keen they were to learn and improve their skills. We would have some fine players available to play when the war was over and we made our way back home to Australia.

A great highlight were 'The Empire Games' where we divided up into our countries of origin and competed against each other. There was nothing better than flogging the Kiwis and the Poms. There were fifteen in our Australian side and everyone had a great time when these events were held. I even won a gold medal in athletics at the games. We also participated in a variety of other activities and sports including, boxing, basketball and soccer whenever the opportunity arose.

From time to time there would be a small group of men awaiting court martial for serious crimes. These crimes included attempts to escape, theft, sabotage and striking a guard. If found guilty they could be sent by the German authorities to a punishment camp in Poland. These Polish camps had appalling reputations. These men were known as Shadow Men and surprisingly they were allowed to mix with the rest of us until their court martial review. Harry Hockey was one of the men in question. He got about the camp under the assumed name of Stanley Hadow or S. Hadow. Not real bright the Huns when it came to understanding our wit.

Beside the delousing hut at the camp there was another small hut that housed our sick and disabled men. The start of an escape tunnel had been built underneath the hut and it was here the Shadow Men would hide until the dreaded Gestapo searches were over and the Gestapo had left the camp. Many of the German guards were friendly with us and also fearful of the Gestapo and if they knew the Gestapo were due at the camp they would let a senior POW officer know. This allowed the senior officer to have the person's name placed on a work detail list showing the person in question was working away from the camp.

Que Sera Sera

On February 18 1945, I was feeling very unwell and presented myself at the camp infirmary. We had two excellent medical orderlies from New Zealand who did a splendid job of keeping us in good health. My complaint resulted in me being admitted to the infirmary for a couple of days.

We knew liberation was getting close; some of the prisoners used ingenious methods of obtaining news on a hidden radio to learn how the war was progressing. This belief was compounded by the frequent presence of Allied aircraft flying over the camp on the way to bomb Germany sites.

I remember the snow had stopped falling during the night but a stiff breeze blew around the camp. Next morning turned out to be a fateful day for the prisoners of Stalag XV111A as American Liberators, looking to bomb the railway close by were unaware the POW camp was in their flight path. They dropped bombs on and near the camp, causing wide spread panic as prisoners ran for cover. The whole camp was totally destroyed.

Another two men and I from the infirmary, we were caught sheltering in one of the open slit trenches right outside the hospital.

A camp in mourning

Australian POW Bill Cassidy from Victoria, remembered coming out from the bomb shelter and realising the men in the slit trench had been buried alive. In his book, *Bill Cassidy: In Love and War*, Barry Cassidy shares his father's recollection of events as they unfolded on that fateful day.

> *'... he remembered that Alan Eason had been admitted to the hospital the day before. Then Bill heard Alan's voice, calm and measured, informing his would-be rescuers from the bottom of the trench that he had been 'badly knocked about.' Alan could just be seen through a gap in the rubble. While the men dug feverishly, he asked for cigarettes to be passed down to him – he said he would be alright if they kept the smokes coming. Two hours later, the men were freed.'*

Alan's mate John Slack was devastated but hopeful as he carried Alan from the trench to the camp hospital but the following day Alan succumbed to his injuries and passed away in the presence of his loyal and devoted mate.

Alan Eason was buried the day after he died on 21 February 1945 in the Klagenfurt War Cemetery, Austria, Plot 2, Row D, Grave 8 along with seven other POWs killed in the same raid.

Those killed from the bombing raid on 19 February 1945 were:

John Lawrence Clarke

Alan Ernest Eason [QLD]

Henry Thomas French

William Jefferson Gilbert [VIC]

Leslie James Howard

Eric Charles Green [QLD]

John Highton

Joseph Melita Riddel [NZ]

The heartbreak and anger experienced by the camp survivors remained with them for many years. Why did these brave men, who had endured so much in the POW camp for four long years, have to die from friendly fire. How devastating for their families back home to learn of this senseless loss of life.

Dreaded telegram

Back in Warwick, the day before the air raid, on 18 February 1945, Alan's mum wrote and posted a letter to him. Alan was never to receive it.

Air Mail

QX17451   *Mrs. C. Eason*
Driver Eason A.E.   *52 Wood St*
Australian Prisoner   *Warwick*
Of War in Germany   *18th Feb*

My Dearest Alan

If you got the last letter that Dad wrote to you I suppose you would be surprised to learn from it that I was in hospital, well here I am home again & not feeling too bad at all. I expect Dad told you the trouble was similar to the operation of last year. The Dr here thinks this will be the end of the trouble. I sincerely hope so, as I am sick of it. Now about yourself Dear. I do hope you are well & keeping that old chin of yours well up & we are hoping it won't be long now. I was so pleased the day of my operation to get a letter from you, it was as good as a tonic especially then as I was feeling pretty awful since then.

I have had three cards also from you. Eleen is here just now, it is Doug's birthday tomorrow & she came over to see him although it is only just over the fortnight since he came. The old hen with only one chick I call her. I am afraid she is very unhappy & looks something awful. Auntie has been good in looking after me & the house while I was away. I wish she could settle down to something. Mrs Cotton is still with us, she has been here for three weeks, but thinks she will be going home on Wednesday with Picots they are going down to the Southport Hotel for three weeks holiday & no doubt he needs it, he is thinner than ever. Their grapes are not any good this year. I had a letter from Raymond this week. He sent me a lovely little brooch a souvenir. I had a letter from Jean the other day & she tells me her silly old Mother is having Turkish baths & gold glint hair washes as Fred the husband likes them slim & pretty. I think she ought to be in a kindergarten school. You say I never mention the new Mrs Eason, well I suppose she is not clever these days she, is very busy preparing for another young Eason that is to arrive in June so when all these young ones arrive, you will have quite a big lot of new relations to meet. You will love Graeme, he will be able to give you cheek. Keith & Clare are going to try & come up for a week-end soon. We picked two nice rock-melons today off our selection. Dad has been very busy this week & had worked all day & is quite happy. It is keeping very dry & hot, it gets like a change every day but passes over. Neville is still with us & is growing tall but is not too fat & can eat like a horse. I had a letter from Moria also one from Vi this week. Did I tell you the Nerang folk told me that Moria is keeping company with Elsie's husband's brother? I do hope she gets a decent husband as she deserves it. Well just at present we are having a storm just now.

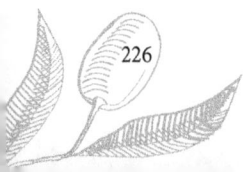

I hope we get a good down pour, it will do a lot of good. Well my Dear this looks like being my issue for this time so will have to draw once more to a close by trusting you are as well as I would like you to be. May God be with you & keep you safe until we meet again. Fondest Love from Dad, Auntie, Eileen & Your ever loving Mother, Florrie Eason.

Ever thinking of you my Dear.

The story of Alan Eason did not end with his death, as John, although missing his mate dreadfully, survived the war and with all the other prisoners was repatriated back to their homelands.

Alan in bed

In his book, *Last of the Light*, published some seventy years later, John wanted Alan Eason's family to know of their friendship in those dark days of the POW Camp and for them to know what a fine man and proud Australian Alan was. John knew his great friend Alan came from a little town called Warwick in Queensland. He knew all about Warwick because in the camp the prisoners would tell stories of their hometowns and childhood experiences.

In the book, John Slack describes the close friendship and the sorrow he felt losing his great friend. Slack had seventy years of built up memories and reverence for his old mate and before he died, he wanted the family to know what a great man Alan was and how he died a hero in that camp. Slack said:

"I lost one of my closest friends in that never to be forgotten day. Alan Eason was a proud and brave Australian. He took the decision to remain above ground that afternoon.

He had taken refuge in a slit trench when the barrage began. That would prove to be a fateful decision. While he did not suffer a direct hit on his person, the impact of the nearby explosion caved in the sides of the slit trench in which he was sheltering... It was winter, the earth was frozen to concrete and he never had a chance... I have never forgotten through the years. I will never forget him. He was one of the finest men you could ever hope to know".

Alan's name shall live forever more on the Roll of Honour boards at Central School and St Andrew's Church (now Uniting Church) in Warwick, Queensland. In addition there is a plaque commemorating his death at Mount Thompson Garden of Remembrance, Brisbane. His name is also registered on the Australian Ex-Prisoners of War Memorial, Ballarat and on the Roll of Honour at the Australian War Memorial, Canberra. Alan's rugby union club, Eagle Junction, had a memorial trophy dedicated to Alan. Teams in the local competition played for the honour of winning this trophy for many years.

AUTHORS NOTE: In April 2020, I was able to make contact with Alan's family. They have been involved in the telling of Alan's story and were very grateful to learn of the book about John Slack written by Alphonsius J Walshe. It is immensely comforting for them to know their Uncle Alan will never be forgotten. For me, it was a great honour to be able to fulfil John Slack's wish of finding Alan's family and passing on John's message.

NATIONAL LIBRARY OF AUSTRALIA

Home Abo

Trove ▼ alan eason 🔍 Adv

🏠 Newspapers / Browse / Warwick Daily News (Qld. : 1919 -1954) / Sat 21 Apr 1945 / Page 4 / Pte. A. Eason Killed in Action

✏ Fix this text 2 📖 ‹ ›

Why are there text errors?

Pte. A. Eason Killed

in Action

Mr. and Mrs C. Eason, Wood
street, received word yesterday
that their youngest son, Private
Alan Eason (27), had been, killed
in action on February 19, 1945. Pte
Eason was taken prisoner of war
at Crete about 4 years ago.

Pte. A. Eason Killed in Action

Mr and Mrs C. Eason, Wood
street, received word yesterday
that their youngest son, Private
Alan Eason (27), had been killed
in action on February 19, 1945. Pte
Eason was taken prisoner of war
at Crete about 4 years ago.

TO-DAY'S EVENTS

CHAPTER 12

Those who stayed

In memory of Albert Edgar Geale

As the author of this book, I believe it is appropriate to conclude this story by honouring and remembering those who stayed. You can imagine the shock I felt when at the very end of my research I discovered one of my own relatives, Albert Edgar Geale, was listed as being buried at the Suda Bay Commonwealth War Cemetery.

I have been able to share my findings and knowledge with my cousins from the Geale family. They in turn have been able to supply me with an extract from a family newsletter published prior to a family reunion held in Yass in 2008.

This is Albert's story.

... Born in 1919 at Temora, the son of William Thompson Geale and Mary Anne Owen, he was a painter at Goulburn when he volunteered on 27 December 1939. He enlisted at Paddington on 27 April 1940 and, after training for four months, he sailed from Sydney and disembarked in Palestine on 30 September 1940. He was in the Allied advance into Libya that captured Bardia and Tobruk. 2/2 Battalion garrisoned Tobruk until 7 March 1941 when it was transferred to Greece, arriving four days later. In a confusing withdrawal the battalion was ordered to hold the Germans at Pineios Gorge with elements of 2/3 battalion and New Zealand forces. They did so for a full day against a German Division of 12,000 men and a Panzer brigade of 250 tanks. Overwhelmed by these forces several parties from 2/2 Battalion managed to join the main withdrawal to the south. Albert was among the last troops to be evacuated from Greece to Crete on 28 April 1941. He, and 188 other members of his battalion, became part of the 16th Australian composite battalion, to help defend Crete from German invasion, which started 20 May 1941. After seven days of fighting the Allied evacuation to Egypt commenced. Albert's battalion was ordered to cover the evacuation at Sfakia. For four nights, between midnight and three o'clock in the morning, warships evacuated 16,000 men – but there was to be no fifth night. Not long after the last ship slipped away the senior officer amongst the remaining 5,000 men, a Lieutenant Colonel, surrendered the force to the Germans. At dawn on 1 June 1941 groups of soldiers stood on the cliffs at Sfakia looking towards the horizon over which the last ships had sailed. News of the surrender had not reached the Luftwaffe and planes strafed the men just after dawn. Albert was killed.

Albert was reported missing on 10 June 1941. But his family did not learn of his death in action until it was confirmed on 25 August 1943.

IN MEMORIUM

Private Albert Edgar Geale, NX13466, 2/2 Infantry Battalion

Killed at Crete. 1 Jun 1941

Buried at Suda Bay War Cemetery

(F. William Thompson)

THOSE WHO SERVED

Source: Geale family papers

The ANZACs

Of the total Commonwealth land force of 32,000 men on Crete, over 18,000 were evacuated, 12,000 were taken prisoner and 2,000 were killed. At this time we must also recognise the 2,000 British sailors who lost their lives in their endeavours to rescue our soldiers in the costly evacuations from Greece and Crete.

The site of Suda Bay War Cemetery was chosen after the war and graves were moved there by 21st and 22nd Australian War Graves Units from the four burial grounds established by the German occupying forces at Chania, Iraklion, Rethymnon and Galata, and from isolated sites and civilian cemeteries.

There are now 1,500 Commonwealth servicemen of the Second World War buried or commemorated in the cemetery. 776 of the burials are unidentified but special memorials commemorate a number of casualties believed to be buried among them. The cemetery also contains 19 First World War burials brought in from Suda Bay Consular Cemetery, one being unidentified. There are also seven burials of other nationalities and 37 non-war burials.

Source: www.cwgc.org

On the next pages is the known list of ANZACs who are buried in the Suda Bay War Cemetery.

Surname	Initials	Service Number
AUSTRALIA		
ADAMS	G R	948583
ADAMS	K L	32194
ADAMS	R	7944
ADDERSON	A A	7751
AGNEW	J J	22651
AIKEN	T J	32209
AITKEN	J E	8269
ALEXANDER	J C G	8877
ALLAN	G A	14214
ALLISON	K A	NX 13467
AMNER	W H	30143
ANDREWS	E M	NX 13351
ANTUNOVICH	F J	29860
APLIN	L T	3214
APPLEBY	E H	7335677
ARCHER	N W B	WX798
ARLIDGE	J T	WX1004
ARMISHAW	A	10148
ARMIT	G N	13667
ARNOLD	L K	5314
ARTHUR	V	33847
ASHMAN	S	3954205
ATKINS	A B	14732
ATOCK	J K	VX5403
ATWELL	L F	VX42541
BAFF	R H	32222
BAIN	W G	2745
BAIRD	J P	NX23840
BAKER	R M	7924
BANISTER	A	NX11170
BANKIER	G M	1629
BARKER	C R	VX 5687
BARNETT	D	28320

Surname	Initials	Service Number
BARNETT	J F	3217
BASSETT	C L	CH/X 100904
BASSETT	J J	30703
BASTABLE	R	7259405
BATES	R E A	NX11170
BAYLY	F P	407402
BEASLEY	F	3532735
BEECHE	N S	25015
BELL	G	1603
BENNETT	C J	2583997
BERRY	K	1463236
BEST	E N	NX3954
BEST	G N	6508
BETTERIDGE	A F	400772
BIRCHALL	V H	4832
BIRD	G G	4204
BIRD	L A	20624
BIRD	T H	10293
BIRTLES	L J	5387
BLACK	R A	7011400
BOLTON	E M W	117444
BOWRON	E E	WX2296
BOYCE	E E E	9294
BOYD	P H A	14201
BRADBURN	D O	1590843
BRADSHAW	K	4861607
BRAMMAR	G D	EX/1708
BREMNER	G R	25501
BRIGGS	N B	6630
BRISTOW	G W	12482
BROADBENT	R E	11051
BROCK	R C A	6916
BROOKSBANK	R O D	206602V
BROWN	A R	60221

Surname	Initials	Service Number
BROWN	C	WX1013
BROWN	T M	3102
BROWN-PRYDE	R B	14229
BROWNE	A R	NX7672
BRYAN	J	63479
BUDDEN	E C	WX317
BULFORD	L V	2696
BURKE	B J	8309
BURNETT	A M S	11589
BURNS	J D	13167
BUTLER	H J	6244
BUTLER	W H M N	NX9920
BYRDE	R G De F	68081
CAMPBELL	E	33928
CAMPBELL	J	1124610
CAMPBELL	R D	22241
CANNON	T C	20587
CAREY	E E	QX571
CAREY	G D	EX/1713
CARR	A	7264188
CARR	G J	14952
CARRICK	R W W	87
CARSON	A E	404537
CARTER	C	22072
CHALMERS	D A	VX12566
CHANDLER	J	1608738
CHAPMAN	R H C	12131
CHARLES	W J	30269
CHISNALL	G E M	4838
CHITTOCK	K A	22842
CHRISTOFI	P	15200
CHURTON	F	10195
CLARK	D	T/159226
CLARK	S T	VX37031

Surname	Initials	Service Number
CLARKE	F L	WX 3337
CLARKE	W G	12975
CLAY	H E	13778
CLYNES	C M	132520
COLBOURN	H A	
COLE	O	20778
COLEMAN	M M	6732
COLEMAN	R F G S	1433769
COLLINGTON	E M	551408
COLLINSON	G H	22261
CONGDON	C C	30051
CONNELL	G W	29518
CONNOR	F	"
COOMBES	G R N	NX14168
COOPER	E F L	3254
COOPER	W E	64140
COPE	E G	6850209
CORNISH	A J	1598836
CORNTHWAITE	L	7596146
CORRIN	S M	1478533
CORSI	W J	3967488
CORY	L E P	2037960
COSTAN	F A	2100061
COURTNEY	W M	39707
COX	E C	401256
COX	G	8629
CRAIG	J	187089
CROSBIE	J	3190340
CROSBY	F A	4745486
CROXFORD	G L	31095
CUMMINS	F J N	NX 11193
CURRY	J	39732
D'AUVERGNE	La T M	7110
DALRYMPLE	T	WX1918

Surname	Initials	Service Number
DALZIEL	O	32365
DANDERSON	J H	4162
DANN	N R	WX503
DAVIES	G T	4600
DAVIES	M G	6023644
DAVITT	J	852518
DE CLIFFORD	T L	7456
DEAN	H A	CHX947
DEVLIN	T J	13238
DEWAR	A R	12701
DIXON	A	NX8183
DOBBIE	R E	10802
DOLAN	W R	NX8152
DONALDSON	D A	405544
DONELLY	H P	7662
DOOLE	W	30413
DOVE	W E	8065
DOWNES	J	NX2812
DOWSETT	A E	WX1869
DRAPER	G J	6023857
DRAYSON	A W B	
DREWERY	R E	14891
DUDLEY	G L R N	NX 2598
DUNCAN	A	11375
DUNCAN	R	7533
DUNCAN	W W	R/115792
DUNN	H E	S/178452
DUNN	W W	2754675
DURRANT	C R	40605
DUTFIELD	F C	4860504
EALAM	E W	8031
EASTON	R T	94326
EDGE	G W	4191187
EDWARDS	G W	PLY/X 1925

Surname	Initials	Service Number
ELLERKER	H V	972580
ELLIOTT	J E	4449749
ELVY	C J	WX842
EPARAIMA	P	39210
ERICKSEN	C G	2824
ERICKSON	W C J	60002
EVANS	R C	NX2813
EVANS	W J	WX2502
EWING	B C	9915
FARDOE	W	7361177
FARRER	R J	EX/563(T)
FAUCKNER	G H	WX1552
FELL	R B	5943
FERGUSON	E J	25143
FERGUSON	T H	3959703
FIELDING	H S	194590
FIRTH	C G	33890
FITZGERALD	M A	WX668
FLAVELL	N A	4995
FLEISHMAN	E D	J/10829
FLETCHER	A T	VX865
FLETCHER	J L	4769
FLOYD	R	12461
FORD	E W	NX7749
FOUNTAIN	L	6485
FRANKLAND	H S	1444726
FRASER	A D	8286
FRASER	J R	WX471
FRIDD	K G	361700
FRIEDLAND	M	96919V
FRIEND	J L	8378
FROELICH	V P	10757
FUSSELL	G K	2618
FYFE	T	9244

Surname	Initials	Service Number
GAGER	W J	QX4034
GALLAGHER	T N	772724
GALYER	N S	32121
GANGELL	S F	WX932
GARBANATI	J L	99818
GEALE	A E	NX13466
GEAR	J R	3203
GEDYE	M H	3205
GEORGE	W L	7218
GIBBONS	J A	627801
GILCHRIST	H	7224
GILES	D B	407498
GILES	D H	33931
GILES	L C	15775
GILES	R A	1152465
GILLAN	A B	549579
GILLANDERS	A S	10214
GILLESPIE	A B	VX12419
GILLIES	R H	12557
GILLIGAN	V W	NX7755
GILMORE	A A	4419
GIRDLESTONE	A	WX419
GIRVEN	T H	653103
GODDARD	J E	406692
GOODHUE	C E	4263
GOODISSON	T R	6035
GORDON	W K	37081
GOSPER	P	NX2448
GOSSE	P E	32492
GOULD	G L	6850332
GOYER	J A	J/85926
GRAFFIN	F	WX2304
GRAHAM	W H	557643
GRAINGER	G K	936229

Surname	Initials	Service Number
GRAY	F H	322609
GREEN	C J	NX6085
GREEN	F W J	WX6115
GREEN	S J	7158
GRIFFIN	J N	VX7124
GRIFFITH	G A	29638
GUEST	J K H	206154V
HACKING	A E	63487
HAGLAND	O	WX903
HAKARAIA	D	39351
HALL	F H	1456163
HALL	G S	127026
HALL	W S	30264
HAMANN	A	
HAMER	D J W	1462289
HAMILTON	J V	100414V
HAMILTON	J	2758306
HAMILTON	J G A	62976
HARDGRAVE	D R	5232
HARDING	W P	39344
HARDY	B	20023
HARRISON	A L	WX737
HARRISON	H F	6896842
HARTREY	E H	1316863
HARVEY	E G	9185
HARWOOD	P E	1264441
HAUGH	G R	1289206
HAWKE	J	VX27838
HAYDEN	W E G	4772
HAZLEWOOD	D P	4856120
HEARN	C B	WX2187
HECKELS	R W	324618
HEMMINGSON	C S	3206
HENLEY	V W J	7602944

Surname	Initials	Service Number
HENSHAW	D J	11084
HEPBURN	J A	10884
HERBERT	L C	7935
HERON	V T	5970
HIGGINS	B E	1514867
HIGGINS	K D	P/MX 65122
HILL	G F	NX11078
HILL	R I	30105
HILL	R E	3791
HIRST	N W	29671
HISLOP	D W	14215
HODGSON	G C	181655
HOFFMAN	A J	8361
HOGAN	H E	10596
HOGGARD	T J	21440
HOLLANDS	P	6847425
HOLLIS	T	29337
HOLMES	C D W	WX2493
HOOPER	T E	654322
HOPLEY	H R H	4324
HORADAM	G I	VX44955
HORLER	J H	13659
HORN	W L C	9394
HORTON	J W	1532340
HOSSACK	W B S	4998
HOTERENE	J	39332
HUBBARD	M H J	20539
HUCKS	J P	322648
HUDSON	J A	NX9923
HUGHES	F	NX13113
HUGHES	T W	1458371
HULME	H C	5381
HUMPHREY	F	4343531
HUMPHREYS	R J	NX15340

Surname	Initials	Service Number
HYDE	A	T/206706
HYMERS	A	551733
IRVING	H	29342
JACKWAYS	A H	3851
JAMIESON	J A	23843
JANECZEK	A	WX2539
JARVILL	F	3709745
JARVIS	J D McD	921009
JELLEY	J	9001
JENNENS	H J	NX2650
JOHN	T	3964746
JOHNS	W R	11070
JOHNSON	F	10844
JOHNSON	R H	3179
JOHNSON	R N	30095
JOHNSTON	A G	3178
JOHNSTON	R	1463566
JOLLY	A C	6896841
JONES	S N	21057
JONES-PARRY	I	170462
JORDAN	A L	2735
JOSSE	K	"
KAIN	B W	7406
KARORA	T	39668
KATENE	T	39368
KEANE	H G D	2598
KEEBLE	E J	5132
KELLY	H F	NX7690
KELLY	J	NX3877
KELLY	T	8857
KEMP	G	6896154
KENYON	S	P/KX 78300
KEREOPA	H	39365
KERR	E H	32293

Surname	Initials	Service Number
KILGOUR	J J H	12872
KING	V B	33192
KINSELLA	F	890193
KIRKMAN	R	5577
KIRKWELL	F H	44926
KNAGGS	A E	1140131
KNIGHT	H A	35736
KNOCKS	S	37444
KNOX	W A	2655907
KYLE	J H S	1486
LAING	G D	25233
LAIRD	C G	3009
LAMONBY	K B	160965
LANDMAN	B G	3184
LANFEAR	W J	29585
LAUDER	T E	29586
LAUZANNE	F J	3050
LAWRENCE	R S	83741
LAWSON	F J	3050
LAYTON	D	VX44655
LEACH	D	39160
LEAVER	L R	852124
LEFROY-OWEN	J	104145
LEITCH	E	NX13132
LEITH	J S	2471
LIDDELL	F	PO/X100215
LINDSAY	W G	29678
LING	E	6896689
LING	R J	23799
LLOYD	F	VX7185
LOADER	E P	885524
LONGHORN	H F	1068456
LONGLEY	F J	6848239
LONGTON	J A	WX1090

Surname	Initials	Service Number
LORD	S	21802
LOWE	H S	11590
LOWE	H N	4861159
LOWRY	C N	29362
LUNDON	J B F	4471
LeCLERC	T R A	NX2921
MACDONALD	J G	33100
MACKENZIE	J D	2066942
MACKIE	E	15055
MACKINDER	W A	31335
MALLOCH	F G	11376
MANSBRIDGE	G	4858852
MAPSTONE	W H	4915368
MARKHAM	J	6008889
MARSHALL	H R	1252
MARTIN	C H	22130
MARTIN	G J	2752739
MARTIN	R H	2590352
MARTIN	R L	29714
MARTIN	W	8799
MASTERS	H K	10848
MATENE	H W	39505
MATHEWS	L E	R/97534
MATHEWS	R J	22996
MATHEWSON	R H	32163
MAXWELL	C J	NX28802
MAXWELL	R J	WX1943
MAY	J H	NX3179
MAYER	R C	WX2059
MAYHEW	L	6847470
MERRICK	A W	NX11956
MERRICK	R V	WX750
MERRYLEES	J W	32771
MEXTED	M F	4477

Surname	Initials	Service Number
MIDDLETON	W	6848076
MILLAR	J A	SX1160
MILLER	J W	WX796
MILLER	R J H	2870
MILLET	E	WX901
MILLIGAN	W E	6681
MILLS	H M	2758473
MILLS	R	CH/X 100508
MILLS	S W	EX/1788
MILNE	C J	NX2454
MILNE	J K	4531
MITCHELL	H F	NX4080
MITCHELL	R W	8514
MONAGHAN	E R	NX7582
MOODY	H D I	WX880
MOON	L W	401011
MOONEY	S L	3193
MOOREY	R C	WX2318
MORAN	J J	4256652
MORIARTY	O B	NX125
MORRIS	J W	159881
MORRIS	R S W	174603
MORRIS	T	T/79471
MORRISON	R	7236
MOSDELL	B H	22689
MOSES	K J	1474
MUIR	R	WX837
MUIRHEAD	W	9265
MUNRO	T M	3053
MURRAY	J	EX1711
MURRAY	J	6976423
MURRAY	K	324585
MacFARLANE	J	3957535
MacGIBBON	N M	30343

Surname	Initials	Service Number
MacLENNAN	J C	WX890
MacLEOD	A	WX5103
MacPHEE	A D K	NX2656
McCALLUM	A R	8312
McCARTHY	W F	20242
McCLEMENTS	J A	3019
McCLYMONT	R B	20054
McCREITH	S	PLY/X 100478
McCULLOCH	H	VX10183
McCURRACH	J H	T/126893
McDAVITT	F	30536
McDERMID	G R W	WX1658
McDONAGH	C W	9213
McDONAGH	W G S	6067
McDONALD	C T	WX1722
McDOWALL	H C	1240
McEWAN	H	12917
McGRATH	F J	31416
McKAY	R	5384
McKENZIE	R A	2706
McKERCHAR	W R	15043
McLEOD	N J	5157
McMINN	D J	6274
McMORLAND	J	643807
McNEILAGE	I G	SX971
McWILLIAMS	T	2754694
NAIRN	R H	2980
NEILL	F H	1482
NEILSON	D H	30091
NELSEY	E O	4795472
NELSON	A	102091V
NELSON	P P	VX616
NEWDICK	C	3058
NEWMAN	K A F	8729

Surname	Initials	Service Number
NEWSON	R H W	NX11063
NEWTON	J L	VX7142
NICHOLSON	A F P	2998
NICHOLSON	J J S	NX7778
NICOLSON	W B	14295
NORTON	A N	6637
O'BRIEN	T E	15874
O'NEILL	J J	6572
OLDHAM	S	8201
OLIPHANT	A E	NX10066
OLSEN	L	WX647
ORME	W G	7593767
OUSLEY	C E J	189964
OWEN	W G	874188
PAGE	F J	1532370
PARATA	D H K M	35998
PARETE	W P B	25976
PARKER	A J	1536811
PARKER	J O	NX3852
PARKER	W D	9164
PARNELL	G F	6606
PARRY	M	3967625
PEARSON	H L	6847749
PEDERSEN	A E	30050
PEEK	I H	25320
PENDLEBURY	J D S	115317
PENNEYSTON	C	293169
PERKINS	D C	1772
PERRETT	G R G	1557836
PERROTT	N M	8058
PETTY	F W	1462236
PICKERING	E V	4353
PIERCE	R H	
PILSBURY	H	1921050

Surname	Initials	Service Number
PIMLOTT	G C	SX1283
PINE	A C	6568
PINNELL	F J	S/75609
PIPER	J J H	83571
PLIMMER	J L R	11716
POA	T R	39197
POLLITT	W	814517
POMEROY	D H	1219525
POPERT	E P	
PORRITT	W J	R/58432
PORTER	R E	6743
POTTS	A	1109273
PRATT	H N	EX/1967
PRENDERGAST	R	14823
PSARA	P A	3887
PURNELL	S A R	4009
QUINN	G F	1543005
RADICH	E	2787
RAHARUHI	R	39437
RANGIPUAWHE	P M	39251
RANKIN	R	15580
RAYMONT	J R	NX11257
REDFERN	F G	2939
REES	H E	"
REID	F J	WX2271
REID	R J	29438
REILLY	C C	40629
RICE	F	207374V
RICE	M C	7191
RICHARDS	L	5217
RICHARDSON	E A	29808
RIDLING	L R	30556
RILEY	E C	NX7711
RIPPINGTON	J S	6849707

Surname	Initials	Service Number
ROBERTS	G F	4032
ROBERTSON	A	551581
ROBERTSON	J C F	2754195
ROBINSON	H L	9697
ROBINSON	R E	NX13200
RODERICK	J J	NX12158
ROGERS	H W	34872
ROLLO	D R	12815
ROSE	G	4441577
ROSS	A M	14688
ROSS	J	6740
ROUSE	J E C	87504
ROWE	W R	20523
ROWNTREE	E T	32989
ROWSWELL	W J	NX4052
RURU	W	26026
RUSSELL	G L	VX15844
RYRIE	H E L	NX5459
SALISBURY	F G	22588
SAMMONS	L W	NX14183
SANDBROOK	K J	6559
SAWYER	F A	945923
SCAIFE	R V	T/64570
SCANDLE	J	29816
SCHMIDT	A	VX6232
SCOLTOCK	H J	7127
SCOTT	F T	8257
SCOTT	H F	2609
SCOTT	R	1454063
SCOTT-WINLOW	N C	104067V
SCOTTHORNE	T A	1527580
SCULLEY	R J	VX13701
SEAMAN	E S	VX5597
SEATON	D E	10787

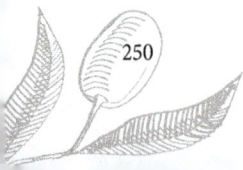

Surname	Initials	Service Number
SEWARD	G	7382073
SHANNON	F R	20457
SHAW	D	8232
SHERBORNE	E W	9477
SHERIDAN	G A M	5502361
SHINER	T E	S/196240
SHIPTON	A	NX9765
SIDAWAY	F	4746562
SIM	G H	8369
SIMS	S V	3956623
SIMS	T	4857929
SIMSON	G V	76097
SINCLAIR	G D	4924
SINNETT	P	1569177
SIRETT	F W	2711
SISARICH	V	5225
SISTERSON	G C	2908
SKILTON	J J	8467
SKIPPER	W	5509
SLADE	W G	6065
SMITH	A F	172941
SMITH	A G	P/J 2609
SMITH	C H	30197
SMITH	G W	12051
SMITH	G A	846558
SMITH	J	PO/X4724
SMITH	N B	9154
SMITH	R D	32088
SMYTH	J W	1870
SNELL	P J	6023794
SNOW	L	4863457
SOMERVILLE	T V	106130
SOUTAR	D	21870
SOUTHWORTH	W J	2611

Surname	Initials	Service Number
SPINLEY	D	3075
SPRING	W B	12296
SPRUNT	S C	558437
SPURDLE	F M	31322
STANSBURY	T G	7340819
STANTON	C	NX7550
STARKIE	W	849633
STEPHENSON	R D	2915
STEWART	H O	4183
STILL	J A E	EX1473
STOCKLEY	A A	29829
STOTT	C H	322688
STRAHORN	A	NX5428
STURGESS	L	9922
SUCKLING	H W	1475908
SULLIVAN	P	35475
SUMMERFIELD	R	WX492
SUTHERLAND	F J	9802
SWEETZIR	W B	PO/21917
SYMMONS	F W	WX739
TABERN	H J	978105
TAKARANGI	H K	6085
TANNAHILL	J	11622
TAPP	W G	6862
TAYLOR	A L	119327
TAYLOR	E P	7518510
TAYLOR	E T	11633
THOMAS	J L	4076566
THOMAS	L	406292
THOMAS	N H	566398
THOMAS	R G	VX4852
THORBJORNSEN	C G	WX683
TIDMARSH	K R	1321966
TIERNEY	D R	VX11698

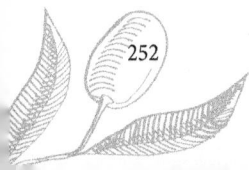

Surname	Initials	Service Number
TILL	C C	1334
TOLLEY	J M	WX1694
TOMS	R A	1875743
TONER	E T H	30225
TOZER	C W	7085
TROJER	J	
TWEEDY	J D	156581
TWOMEY	R E	1603221
UNDERHAY	S W	15198
URQUHART	R S	36987
UTTING	J P	WX823
VALE	F C	VX1160
VAN ASCH	J F	1285
VEAL	H G	WX2276
VELGOLASKY	L	31988
VERCOE	H T R	34877
VERDON	L A	10630
VESTY	R L	37428
VINCENT	E	3967485
VOKES	P H	3840
WADDELL	T R	3090
WADDINGTON	R	6768
WAGNER	C	
WAGNER	R	PAL/13601
WAKELIN	N L	6265
WALDECK	J K	WX726
WALDEN	F	4861712
WALKER	C W	T/16691
WALKER	J E	26238
WALL	L	31988
WALLEN	C M	2927
WALSH	J P	22421
WALTERS	E W C	551094
WARD	J W	12461

Surname	Initials	Service Number
WARDROP	D	133920
WARMINGTON	E S	21885
WARNER	E G	25422
WATERMEYER	N M	207152V
WATKINS	L J	3959695
WATSON	E	2656593
WATSON	F	2754395
WATSON	L A	13626
WATSON	R W	8164
WEBB	E J	NX4116
WEBB	S A	NX7802
WEBBER	C	621163
WEBSTER	W	6833
WEDGWOOD	K S	401260
WEHI	M P	39802
WEIR	R D H	WX2408
WELLINGTON	R A	7896
WELLS	H F	94436V
WEST	G P	117320
WESTWOOD	T C	1190196
WHETTON	K	7518455
WHITTEKER	D N	15385
WHORLOW	R G	873092
WILDASH	R F	16161
WILKINS	D B	WX2014
WILKINS	J S	3832
WILLCOCKS	W A	556474
WILLIAMS	E J	3957130
WILLIAMS	F A	22563
WILLIAMS	G J	10522
WILLIAMS	P A	39728
WILLIAMS	R E	30087
WILLIAMSON	A D	30012
WILLIAMSON	C F	

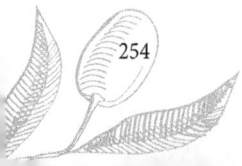

Surname	Initials	Service Number
WILLIS	A G	30113
WILSON	C	7265
WILSON	D J	VX5538
WILSON	H	1512796
WILSON	L A	903847
WILSON	R C	10655
WINDYBANK	D O	2972
WINTER	C L	13693
WITHERS	J E	NX5974
WOOD	F T	NX3883
WOOD	J A D	7347432
WOODLAND	W F	NX12793
WOOTTON	R J S	42088
WORT	J H	3966328
WRIGHT	J	88664
WRIGLEY	B	5256
YOUNG	A	WX857
YOUNG	W G	156331
ZBYSZYNSKI	J	P/0157
NEW ZEALAND		
ADAMS	K L	32194
ADAMS	R	7944
ADDERSON	A A	7751
AGNEW	J J	22651
AIKEN	T J	32209
AITKEN	J E	8269
ALEXANDER	J C G	8877
ALLAN	G A	14214
AMNER	W H	30143
ANTUNOVICH	F J	29860
APLIN	L T	3214
ARMISHAW	A	10148
ARMIT	G N	13667

Surname	Initials	Service Number
ARNOLD	L K	5314
ARTHUR	V	33847
ATKINS	A B	14732
BAFF	R H	32222
BAIN	W G	2745
BAKER	R M	7924
BANKIER	G M	1629
BARNETT	J F	3217
BASSETT	J J	30703
BEECHE	N S	25015
BELL	G	1603
BEST	G N	6508
BIRCHALL	V H	4832
BIRD	G G	4204
BIRD	L A	20624
BIRD	T H	10293
BIRTLES	L J	5387
BOYCE	E E E	9294
BOYD	P H A	14201
BREMNER	G R	25501
BRIGGS	N B	6630
BRISTOW	G W	12482
BROADBENT	R E	11051
BROCK	R C A	6916
BROWN	A R	60221
BROWN	T M	3102
BROWN-PRYDE	R B	14229
BULFORD	L V	2696
BURKE	B J	8309
BURNETT	A M S	11589
BURNS	J D	13167
BUTLER	H J	6244
CAMPBELL	E	33928
CAMPBELL	R D	22241

Surname	Initials	Service Number
CANNON	T C	20587
CARR	G J	14952
CARTER	C	22072
CHAPMAN	R H C	12131
CHARLES	W J	30269
CHISNALL	G E M	4838
CHITTOCK	K A	22842
CHURTON	F	10195
CLARKE	W G	12975
CLAY	H E	13778
COLE	O	20778
COLEMAN	M M	6732
COLLINSON	G H	22261
CONGDON	C C	30051
CONNELL	G W	29518
COOPER	E F L	3254
COURTNEY	W M	39707
COX	E C	401256
COX	G	8629
CROXFORD	G L	31095
CURRY	J	39732
D'AUVERGNE	La T M	7110
DALZIEL	O	32365
DANDERSON	J H	4162
DAVIES	G T	4600
DE CLIFFORD	T L	7456
DEVLIN	T J	13238
DEWAR	A R	12701
DOBBIE	R E	10802
DONELLY	H P	7662
DOOLE	W	30413
DOVE	W E	8065
DREWERY	R E	14891
DUNCAN	A	11375

Surname	Initials	Service Number
DUNCAN	R	7533
DURRANT	C R	40605
EALAM	E W	8031
EPARAIMA	P	39210
ERICKSEN	C G	2824
ERICKSON	W C J	60002
EWING	B C	9915
FELL	R B	5943
FERGUSON	E J	25143
FIRTH	C G	33890
FLAVELL	N A	4995
FLETCHER	J L	4769
FLOYD	R	12461
FRASER	A D	8286
FRIEND	J L	8378
FROELICH	V P	10757
FUSSELL	G K	2618
FYFE	T	9244
GALYER	N S	32121
GEAR	J R	3203
GEDYE	M H	3205
GEORGE	W L	7218
GILCHRIST	H	7224
GILES	D H	33931
GILES	L C	15775
GILLANDERS	A S	10214
GILLIES	R H	12557
GILMORE	A A	4419
GOODHUE	C E	4263
GOODISSON	T R	6035
GORDON	W K	37081
GOSSE	P E	32492
GREEN	S J	7158
GRIFFITH	G A	29638

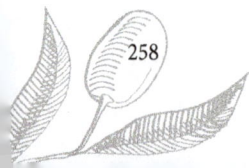

Surname	Initials	Service Number
HAKARAIA	D	39351
HALL	W S	30264
HARDGRAVE	D R	5232
HARDING	W P	39344
HARDY	B	20023
HARVEY	E G	9185
HAYDEN	W E G	4772
HEMMINGSON	C S	3206
HENSHAW	D J	11084
HEPBURN	J A	10884
HERBERT	L C	7935
HERON	V T	5970
HILL	R I	30105
HILL	R E	3791
HIRST	N W	29671
HISLOP	D W	14215
HOFFMAN	A J	8361
HOGAN	H E	10596
HOGGARD	T J	21440
HOLLIS	T	29337
HOPLEY	H R H	4324
HORLER	J H	13659
HORN	W L C	9394
HOSSACK	W B S	4998
HOTERENE	J	39332
HUBBARD	M H J	20539
HULME	H C	5381
IRVING	H	29342
JACKWAYS	A H	3851
JAMIESON	J A	23843
JELLEY	J	9001
JOHNS	W R	11070
JOHNSON	F	10844
JOHNSON	R H	3179

Surname	Initials	Service Number
JOHNSON	R N	30095
JOHNSTON	A G	3178
JONES	S N	21057
JORDAN	A L	2735
KAIN	B W	7406
KARORA	T	39668
KATENE	T	39368
KEANE	H G D	2598
KEEBLE	E J	5132
KELLY	T	8857
KEREOPA	H	39365
KERR	E H	32293
KILGOUR	J J H	12872
KING	V B	33192
KIRKMAN	R	5577
KNIGHT	H A	35736
KNOCKS	S	37444
KYLE	J H S	1486
LAING	G D	25233
LAIRD	C G	3009
LANDMAN	B G	3184
LANFEAR	W J	29585
LAUDER	T E	29586
LAUZANNE	F J	3050
LEACH	D	39160
LEITH	J S	2471
LINDSAY	W G	29678
LING	R J	23799
LORD	S	21802
LOWE	H S	11590
LOWRY	C N	29362
LUNDON	J B F	4471
MACDONALD	J G	33100
MACKIE	E	15055

Surname	Initials	Service Number
MACKINDER	W A	31335
MALLOCH	F G	11376
MARSHALL	H R	1252
MARTIN	C H	22130
MARTIN	R L	29714
MARTIN	W	8799
MASTERS	H K	10848
MATENE	H W	39505
MATHEWS	R J	22996
MATHEWSON	R H	32163
MERRYLEES	J W	32771
MEXTED	M F	4477
MILLER	R J H	2870
MILLIGAN	W E	6681
MILNE	J K	4531
MITCHELL	R W	8514
MOONEY	S L	3193
MORRISON	R	7236
MOSDELL	B H	22689
MOSES	K J	1474
MUIRHEAD	W	9265
MUNRO	T M	3053
MacGIBBON	N M	30343
McCALLUM	A R	8312
McCARTHY	W F	20242
McCLEMENTS	J A	3019
McCLYMONT	R B	20054
McDAVITT	F	30536
McDONAGH	C W	9213
McDONAGH	W G S	6067
McDOWALL	H C	1240
McEWAN	H	12917
McGRATH	F J	31416
McKAY	R	5384

Surname	Initials	Service Number
McKENZIE	R A	2706
McKERCHAR	W R	15043
McLEOD	N J	5157
McMINN	D J	6274
NAIRN	R H	2980
NEILL	F H	1482
NEILSON	D H	30091
NEWDICK	C	3058
NEWMAN	K A F	8729
NICHOLSON	A F P	2998
NICOLSON	W B	14295
NORTON	A N	6637
O'BRIEN	T E	15874
O'NEILL	J J	6572
OLDHAM	S	8201
PARATA	D H K M	35998
PARETE	W P B	25976
PARKER	W D	9164
PARNELL	G F	6606
PEDERSEN	A E	30050
PEEK	I H	25320
PERKINS	D C	1772
PERROTT	N M	8058
PICKERING	E V	4353
PINE	A C	6568
PLIMMER	J L R	11716
POA	T R	39197
PORTER	R E	6743
PRENDERGAST	R	14823
PURNELL	S A R	4009
RADICH	E	2787
RAHARUHI	R	39437
RANGIPUAWHE	P M	39251
RANKIN	R	15580

Surname	Initials	Service Number
REDFERN	F G	2939
REID	R J	29438
REILLY	C C	40629
RICE	M C	7191
RICHARDS	L	5217
RICHARDSON	E A	29808
RIDLING	L R	30556
ROBERTS	G F	4032
ROBINSON	H L	9697
ROGERS	H W	34872
ROLLO	D R	12815
ROSS	A M	14688
ROSS	J	6740
ROWE	W R	20523
ROWNTREE	E T	32989
RURU	W	26026
SALISBURY	F G	22588
SANDBROOK	K J	6559
SCANDLE	J	29816
SCOLTOCK	H J	7127
SCOTT	F T	8257
SCOTT	H F	2609
SEATON	D E	10787
SHANNON	F R	20457
SHAW	D	8232
SHERBORNE	E W	9477
SIM	G H	8369
SINCLAIR	G D	4924
SIRETT	F W	2711
SISARICH	V	5225
SISTERSON	G C	2908
SKILTON	J J	8467
SKIPPER	W	5509
SLADE	W G	6065

Surname	Initials	Service Number
SMITH	C H	30197
SMITH	G W	12051
SMITH	N B	9154
SMITH	R D	32088
SMYTH	J W	1870
SOUTAR	D	21870
SOUTHWORTH	W J	2611
SPINLEY	D	3075
SPRING	W B	12296
SPURDLE	F M	31322
STEPHENSON	R D	2915
STEWART	H O	4183
STOCKLEY	A A	29829
STURGESS	L	9922
SULLIVAN	P	35475
SUTHERLAND	F J	9802
TAKARANGI	H K	6085
TANNAHILL	J	11622
TAPP	W G	6862
TAYLOR	E T	11633
TILL	C C	1334
TONER	E T H	30225
TOZER	C W	7085
UNDERHAY	S W	15198
URQUHART	R S	36987
VAN ASCH	J F	1285
VELGOLASKY	L	31988
VERCOE	H T R	34877
VERDON	L A	10630
VESTY	R L	37428
VOKES	P H	3840
WADDELL	T R	3090
WADDINGTON	R	6768
WAKELIN	N L	6265

Surname	Initials	Service Number
WALKER	J E	26238
WALLEN	C M	2927
WALSH	J P	22421
WARMINGTON	E S	21885
WARNER	E G	25422
WATSON	L A	13626
WATSON	R W	8164
WEBSTER	W	6833
WEHI	M P	39802
WELLINGTON	R A	7896
WHITTEKER	D N	15385
WILDASH	R F	16161
WILKINS	J S	3832
WILLIAMS	F A	22563
WILLIAMS	G J	10522
WILLIAMS	P A	39728
WILLIAMS	R E	30087
WILLIAMSON	A D	30012
WILLIS	A G	30113
WILSON	C	7265
WILSON	R C	10655
WINDYBANK	D O	2972
WINTER	C L	13693
WRIGLEY	B	5256

Suda Bay War Cemetery

The Geale and Wheeler families wish to pay our respects to Albert and all those who fought in this conflict.

In closing I would like to finish this work with the haunting words of Rupert McCall's poem *The Reason Why* in tribute to these men.

THE REASON WHY

Find me there when my heart lies bare
And lift my soul to the sky
Let me soar from this field of war
With the birds that freely fly

Clear the ground as I look back down
Of the blood and the blast and the hurt
Allow my eyes to replace those cries
With beams that rise from the dirt

Four strong beams that define the dreams
Of the sunburnt land I love
Four proud stones that revive my bones
And salute the ghosts above

Courage first - it goes unrehearsed
When the field explodes with fire
To confront your fear, then to persevere
Is to epically inspire

Endurance next and the spirit flexed
In the heat of that endless track
To push though pain where our men remain
A Digger won't step back

And if one man falls, then Mateship calls
It's a non-negotiable trait
At the soldier's core, no bond means more
Than to stand beside a mate

Which adds to the sting of the hardest thing
In the sad and silent strain
The terrible cost of a good mate lost
The Sacrifice never in vain

The sacrifice never forgotten!
The tragedy never disguised
The legacy bravely defended!
The freedom eternally prized!

The son of a mother and father
A hero my country will claim
The strength to walk forward together
Our future to honour their name

Yes find me there when my heart lies bare
And lift my soul to the sky
Allow my day to find peace this way
…and to know the reason why…

Rupert McCall ©

LEST WE FORGET

EPILOGUE

In hindsight, history enlightens us to why our soldiers were ever involved in this failed conflict. It shows that while this campaign was a failure, in the long run our soldiers were outstanding in the execution of their duties, and their efforts were valuable in the overall victory of the Allies against the Axis powers.

Of the 58,000 troops sent to Greece, some 35,000 of the fighting formations were made up of a division each from Australia and New Zealand and one armoured brigade from Britain. The balance consisting of administration staff, communications and supply staff, medical and non-fighting personnel.

It had been hoped the Yugoslavians would be able to hold the border on their side and our troops would be able to hold the passes in Greece. The Germans quickly dispelled any hope of this. The sheer number of their tanks (800 to our 100) was a fair indication of the one-sidedness of the battle.

But the Germans suffered as heavy causalities here as anywhere in the whole campaign, thanks to the heroic efforts of our machine gun battalion, artillery regiment, anti-tank regiment and the armoured brigades.

Conditions were impossible for the RAF fighters as they were severely outnumbered and could offer little protection for the Allied troops. Tragically their losses during this period were horrendous.

Some of the stories recalled by the survivors included how one Australian unit continued to fight until they were 20 miles behind German lines before escaping along a stream. This stubborn rear-guard action gained precious time for their fellow troops to retreat. At Larissa, ANZACS were trapped and encircled by the enemy, but managed to fight their way out and escape. Many thousands of the Allied troops were safely evacuated to Alexandria and Crete.

Of the 40,000 troops who were on the island of Crete, approximately one quarter were fresh troops; the balance was made up of battle-weary soldiers with few or no weapons who had been evacuated from Greece, and non-combatant troops.

Written and oral accounts state quite simply that when the Greek Government conceded defeat they honourably thanked their Allies for their help, acknowledging the Allies had done everything in their power to save Greece from invasion. Before Greece was forced to capitulate, they recognised the war was not lost and encouraged the Allies to save themselves and continue their efforts so they could go on to help win the war elsewhere. Wise words, as this is exactly what happened. While the battles in Greece and Crete were lost, the Allies did go on to win the war.

In the words of Winston Churchill:

> 'This is a war of peoples and of causes ... whose names will never be known, whose deeds will never be recorded. This is a war of the unknown warriors...'

It is only fitting at this stage to record the fate of two of the most notorious people involved in the Second World War—Adolf Hitler and the infamous traitor, Lord Haw-Haw.

Adolf Hitler was the man who wanted to rule the world at any cost. He was driven by greed and his own over-inflated sense of self-importance. He was someone to whom the loss of human life in order to achieve his ultimate goal held no importance. How quickly do men of this ilk fall when events go against them and their visions of total control and leadership disappear in defeat? Going from the heights of invincibility to the depths of destruction is inconceivable to them.

Hitler could see his dream and single-minded determination of controlling the world at any cost crumble around him. By April 1945, the Allies had gained the upper hand and had taken control of the war, defeating the Axis powers on all fronts.

Adolf Hitler, like other dictators before him, took the coward's way out. Hiding in a bunker, broken and crushed mentally, he opted to commit suicide rather than be captured and answer for the horrendous crimes against humanity he had orchestrated.

As fate would dictate, it was on this same day, 30 April 1945, in Hamburg where William Joyce made his last broadcast as Lord Haw-Haw.

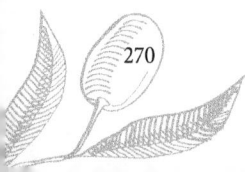

Earlier in 1932 Joyce had joined the British Union of Fascists and was appointed director of propaganda because of his great oratory. However his bad temper and willingness to brawl with anyone who did not agree with his outlook led to him being sacked from the party. Furious with this turn of events he went on to form his own organisation called the 'Nationalist Fascist League'. This organisation was a failure and in August 1939 Joyce and his wife fled to Germany.

He auditioned for German radio and was hired immediately and became the best-known English propagandist and inherited the pseudonym, 'Lord Haw-Haw'. His broadcasts were so effective Adolf Hitler awarded Joyce the War Merit Cross (First and Second Class).

His reign on the airwaves came to an abrupt end when Hitler committed suicide and Germany surrendered to Allied forces a week later. Joyce realised it was time to get out of Germany before he was captured and taken to task by the Allies. Joyce tried escaping to Denmark but failed to reach his objective. He and his wife were captured by two British officers who found them hiding near the border. Joyce was transported to Britain and tried for treason. Pronounced guilty at his trial, he was executed at Wandsworth Prison on 3 January 1946.

AFTERWORD

As for those who returned home, many people expected them to forget and go on with life as before. Life did go on but for most service personnel their wartime experiences changed the way they saw things in the future. Like every other returned serviceman and woman who has been involved in active duty on a war front, things can never be the same again. For many veterans it gave them a greater understanding and appreciation of what their forefathers had experienced in the First World War. It also shed clarity ,to a greater degree, on why those men and women lived the lives they did when they returned from the battle front.

Many chose not to speak about the things experienced unless it was with old mates. Some wrote books to help expunge the experiences and horrors that befell them and their fellow unit members. Others would only talk about their experiences if they could make a joke of it, this was particularly true for Douglas Channell MC. Having said that, it was something many would have done all over again if the need to defend families and country had arisen again in their life time.

Now you have finished reading this book I hope it has inspired you to look into your own family history and see what stories you can uncover for yourself.

The stories contained in this book have been written to share how these men felt and lived during this brief period in our history. It highlights how we can commemorate people in our own way using their own words, even though they may be long gone.

Being given access to diaries, letters, newspaper clippings and being able to converse with members of these men's families has allowed me to capture in some small way the essence of these men.

Writing this book along the lines of a fictional work based in historical fact allowed me to meld stories together giving it a richer context rather than a straight report of facts. The writing was intended to give the reader the sense of travel, adventure and hardships experienced by men of different ranks and units who fought side-by-side to accomplish the same outcome. Unfortunately bungled communications combined with the strength of the enemy, meant victory was not to be during this short and little recorded event in our military history.

Like most Veterans, these men wonder why someone would want to know their story; to them, they were just like everyone else. Those of us not involved in these conflicts have the freedom to look back and see how their endurance, courage, humour, resourcefulness, mateship, faith and devotion to duty and to each other have given us the freedoms we enjoy today.

If you are interested in learning more please reach out and contact me online at *www.deborahcwheeler.com*

ACKNOWLEDGEMENT

How do you convey your appreciation to someone who has opened up a whole new world for you? Andreas, taking the time to stop and speak with me on that autumn morning by the side of the road has allowed me to not only expand my knowledge of this slice of our military history but connect with some wonderful descendants of the men I have written about. Your passion and dedication to your cause was so inspiring. Thank you for allowing me to walk alongside you on this part of your life's journey.

The highlight of researching and writing this book came when I was given a single name, Alf Carpenter. It is not often you get the rare privilege of speaking personally with someone who was involved in a conflict that took place 80 years ago. At 103 years of age, Alf has been an inspiration and a master of information. Thank you Alf from the bottom of my heart.

I would also like to acknowledge the amazing families of the men who have been portrayed in my book; I could not have done this without you. Andreas, Roxane and Stuart Scott, Graeme and Sandy Eason, Lyn Collins, Lisa Zampelis, Lesley Mills, Jennifer Lister, David Morris, Glenda Humes, Dorothy Burton and Judith Standen.

John Telfer, thanks for sharing your knowledge on Alan Eason. Because of your generosity, I have been able to complete a task for the late veteran, John Slack. John's wish to get his message to Alan's family has now been accomplished some 75 years after Alan's death.

To those who shared their works with me and granted me permission to use small snippets of those works in this book, thank-you.

They say life is what you make of it. I am privileged to have found something I love doing and sharing with others. Thank you to everyone who had faith in me when I didn't believe in myself. For my good friend Jane who had the courage to speak up and let me know I had dyslexia. Now I can work around the cause of my frustration of knowing what I wanted to say but not being able to get it down on paper.

I look forward to hearing from you, my readers. Your interaction with me gives me the inspiration and strength to keep going when at times the way seems all uphill.

Deborah

BIBLIOGRAPHY

Books

BIANK, Major Maria A. *The Battle of Crete: Hitler's Airborne Gamble.* Pickle Partners Publishing, 2014.

BORRIE, John. *Despite Captivity.* London: Kimber, 1975.

CASSIDY, Barrie. *Private Bill: In Love and War.* Melbourne: Melbourne University Publishing, 2014.

HETHERINGTON, John. *Airborne Invasion: The Story of the Battle of Crete.* Sydney: Angus and Robertson, 1944.

HOLMES, Peter. Crete 1941 - *The Road to Sphakia: An epic in the Battle for Crete, May 1941.* Reedy Creek, Queensland : Peter Lloyd Holmes Publication, 2015.

LINDSAY, Jack. *Beyond Terror: a novel of the Battle of Crete.* London: Andrew Dakers, 1943.

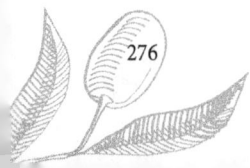

MONSON, Ronald. *The Battle of Greece*. Supplement to South Eastern District RSL DIGEST – Vol. II, No 6, April 1991.

MONTEATH, Peter. *Battle on 42nd Street: War in Crete and the Anzacs' bloody last stand*. Kensington, NSW: NewSouth Books, 2019.

MacDONALD, Callum. *The Lost Battle: Crete*, 1941. London: Macmillan, 1993.

SMITH, Kelsey A. *The Balance Sheet of the Battle of Crete: How Allied Indecision, Bureaucracy, and Pretentiousness Lost the Battle*. Marine Corps Command And Staff Coll Quantico VA (Corporate Author); United States Defense Technical Information Center.

WALSHE, Alphonsius John. *The Last of the Light*, EBooks by Design, 2013.

Newsletters

2/1 Tank Attack regiment Association Royal Australian Artillery, 1997-2001 - Complied by Alan Grant QX351.

Website links

"BRITAIN DECLARES WAR" *The Courier-Mail (Brisbane, Qld.: 1933 - 1954)* 4 September 1939: 1. Web. 30 Jul 2020
http://www.nla.gov.au/nla.news-article40908629

"Jobs' At Home For Men Of 2nd A.I.F." *The Courier-Mail (Brisbane, Qld : 1933 - 1954)* 20 March 1940: 5. Web. 30 Jul 2020
http://nla.gov.au/nla.news-article40860536

"THE MAN IN THE STREET" *Smith's Weekly (Sydney, NSW: 1919 - 1950)* 27 September 1941: 7. Web. 14 Aug 2020
http://nla.gov.au/nla.news-article234602882

"Retimo—A Lost Victory" *The Sydney Morning Herald (NSW: 1842 - 1954)* 1 June 1946: 6. Web. 30 Jul 2020
http://nla.gov.au/nla.news-article17981840

Service Records: National Archives of Australia
https://www.naa.gov.au

Stalag 18a by Ian Brown
www.stalag18a.org

Prime Minister Robert G. Menzies: wartime broadcast. AWM
https://www.awm.gov.au/articles/encyclopedia/prime_ministers/menzies

Virtual War Memorial Australia
https://www.vwma.org.au

Online Cenotaph – Auckland War Memorial Museum
https://www.aucklandmuseum.com>war-memorial>online-cenotaph

Veterans organisations

42 for 42 www.42for42.org.au

PTSD: National Center for PTSD Home - Veterans Affairs
www.ptsd.va.gov

Soldier On Australia
soldieron.org.au

Veterans Assistance | RSL DefenceCare
www.defencecare.org.au › services › services

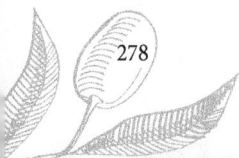

GLOSSARY

Bivouac – A temporary camp without tents or cover

Bloke – Man

Chin wag – Conversation

Gratis – Free

HQ – Headquarters

Huns - Germans

Lustre Force – Code name for NZ Forces in Greece and Crete

M&M Stew – Stew containing meat and vegetables

Man of Confidence – Person nominated from within the ranks of POWs who would act in the interest of all POWs.

MC – Military Cross

MBE – Member of the British Empire

MID – Mentioned in Dispatches

Orderly Room – a room in barracks sometimes occupied by the first sergeant that contains the company, troop, or battery records and is used for company business.

Pickets – a group of soldiers detailed for a specific duty

POW – Prisoner of War

R.A.P. – Regimental Aid Post

Rag tag – Miscellaneous

RSM – Regimental Sergeant Major

Strafing – To bombard with constant machine gun fire

X List – Recorded personnel who were absent from their regular units for one reason or another.

BIOGRAPHY

Deborah Wheeler lives in Warwick, Queensland with her husband Ross Price. She and her husband run a traditional Bed & Breakfast in their home, a beautifully restored Queenslander built in 1902. Deborah often entertains guests when conducting author talks and tours covering many topics including aspects of the rich local military history that abounds in the region.

After qualifying as a library technician in 1990, she worked in a variety of libraries for over 20 years.

She also worked in all aspects of newspaper publishing, and travelled around the east coast of Australia as an advertising consultant.

In 2006, while working as manager of a small bowls club in Queensland, she was faced with the challenge of finding information about other clubs for travelling bowlers.

There was no relevant information available, so drawing on her grounding in library, information services and advertising, Deborah took the gigantic step of entering the world of publishing; producing the first and only national lawn bowls directory in Australia. Fifteen years later, it is still thriving and has progressed to its current online format including a free mobile app for bowlers.

Writing and research became her passion after being commissioned in 2015 to undertake a research project for the Stanthorpe RSL Sub Branch on the long forgotten military medical facility, the Kyoomba Sanatorium. Uncovering stories of 550 soldiers, doctors, nurses and others involved at the sanatorium became a light bulb moment for her. An accidental historian was born.

Since writing and self-publishing a series of five broadsheets based on the Kyoomba Sanatorium, she has gone on to publish *Kyoomba Sanatorium 1916-1935 Volumes 1 & 2*, *Tales of a Military Medal Recipient* and a *Lancaster Bomber Rear Gunner*, and *My Pop Was a Kangaroo ANZAC*. All of these titles are held in the prestigious Australian War Memorial collection.

Since then Deborah has been commissioned to write six books for various individuals and groups. She also publishes books for other writers and is proud to acknowledge one of these works by a close friend was also accepted into the Australian War Memorials Collection alongside her own works.

Deborah shared part of her own family's military history when she produced a touching story about her grandfather William James Wheeler, which is told in loving detail using vivid illustrations to teach our current and future generations why the ANZAC story is so important to us to this day.

In September 2019, Deborah was invited to be the Keynote Speaker at the National Archives of Australia, Brisbane Office monthly addresses. Her presentation was the only non-Archive staff presentation for the year.

Deborah enjoys meeting people from all walks of life, travelling and paying forward.

With five works held in the Australian War Memorial collection, Deborah encourages you to connect with her and share information about your relatives who fought in the AIF during the First and Second World Wars. Deborah collects information and stories about these men for a national database of veterans. Her mission is to uncover as many stories and facts as possible about our ANZACs particularly those connected with the First World War.

Email: read@deborahcwheeler.com

Web: www.deborahcwheeler.com

www.ingramcontent.com/pod-product-compliance
Lightning Source LLC
Chambersburg PA
CBHW071548210326
41597CB00019B/3162